essentials

Springer essentials provide up-to-date knowledge in a concentrated form. They aim to deliver the essence of what counts as "state-of-the-art" in the current academic discussion or in practice. With their quick, uncomplicated and comprehensible information, essentials provide:

- an introduction to a current issue within your field of expertise
- an introduction to a new topic of interest
- an insight, in order to be able to join in the discussion on a particular topic

Available in electronic and printed format, the books present expert knowledge from Springer specialist authors in a compact form. They are particularly suitable for use as eBooks on tablet PCs, eBook readers and smartphones.

Springer essentials form modules of knowledge from the areas economics, social sciences and humanities, technology and natural sciences, as well as from medicine, psychology and health professions, written by renowned Springer-authors across many disciplines.

Susanne Schindler-Tschirner
Werner Schindler

Mathematical Stories II – Recursion, Divisibility and Proofs

For Gifted Students in Primary School

 Springer

Susanne Schindler-Tschirner
Sinzig, Germany

Werner Schindler
Sinzig, Germany

ISSN 2197-6708   ISSN 2197-6716   (electronic)
essentials
ISBN 978-3-658-38610-8   ISBN 978-3-658-38611-5   (eBook)
https://doi.org/10.1007/978-3-658-38611-5

This Springer imprint is published by the registered company Springer Fachmedien Wiesbaden
GmbH, part of Springer Nature.
The registered company address is: Abraham-Lincoln-Str. 46, 65189 Wiesbaden, Germany

What You Can Find in This essential

- Learning units in stories
- Gaussian summation formula and a recursion formula
- Elementary combinatorics
- Divisibility, prime factorization, modular arithmetic (modulo calculation)
- Proofs
- Sample solutions

Preface

The conception and design of this *essentials* and of Volume I of the "Mathematical Stories" (Schindler-Tschirner and Schindler 2019) are the result of the experience gained from a mathematics study group for gifted students, which the second author led at the primary school in Oberwinter (Rhineland-Palatinate). Twelve students from grades 3 and 4 took part. This was 10% of all students in these two grades. Of these, at least[1] three students were later able to win prizes in national mathematics competitions.

Of course, the authors do not assume that these successes were only made possible by participation in this mathematics study group. Rather, with the two *essentials*, we would like to make a contribution to awakening interest in and enjoyment of mathematics and promoting mathematical talent.

Sinzig, Germany
January 2019

Susanne Schindler-Tschirner
Werner Schindler

---

[1] The authors are no longer in contact with all participants of the mathematics study group.

Contents

Introduction

1

Like the first volume of "Mathematical Stories" (Schindler-Tschirner & Schindler, 2019), this *essential* consists of two parts. Again, Part I contains six chapters of tasks and Part II contains the sample solutions discussed in detail with didactic suggestions, mathematical goals, and outlooks. The tasks are integrated into a narrative that continues the first volume.

Like the first volume (Schindler-Tschirner & Schindler, 2019), this *essential* is aimed at leaders of study groups (SGs) and remedial courses for mathematically gifted students in grades 3 and 4, at teachers who practice differentiated mathematics lessons, and also at committed parents for out-of-school support. The sample solutions are tailored to the management of SGs; modified accordingly, however, they can also serve as a guide for parents who work through this book together with their children.

Although both volumes are essentially self-contained, it is recommended that you begin with Volume I.

1.1 Mathematical Goals

The "Mathematical Stories" differ fundamentally from some pure task collections, which contain interesting and by no means trivial mathematics tasks "to puzzle with," but in which, in our view, fall short of targeted learning and application of mathematical techniques. As in Volume I, the goal of this *essential* is to teach students basic mathematical techniques and to awaken enjoyment of mathematics.

© The Author(s), under exclusive license to Springer Fachmedien
Wiesbaden GmbH, part of Springer Nature 2023
S. Schindler-Tschirner, W. Schindler, *Mathematical Stories II - Recursion,
Divisibility and Proofs*, essentials,
https://doi.org/10.1007/978-3-658-38611-5_1

Like Volume I, this *essential* does not elaborate on general didactic considerations and theories of gifted education, although the bibliography contains a selec-
tion of relevant publications for the interested reader. This *essential* concentrates
on the tasks, the mathematical methods and techniques applied, and on concrete
didactic suggestions for implementation in a gifted SG. This second volume also
contains tasks that have hardly any comparable examples in normal school lessons
and that challenge and encourage children's mathematical thinking. The tasks also
represent a new challenge for gifted students. Unlike in normal school lessons, they
need a lot of perseverance here to be able to solve the tasks set. A special mathematics textbook in primary school is not required.

As in the first volume of the "Mathematical Stories", the students are guided by
the tasks to work out the solutions as independently as possible (but probably with
specific help from the course instructor!). Solving the tasks again requires a high
degree of mathematical imagination and creativity; qualities that are fostered by
dealing with mathematical problems.

The first volume (Schindler-Tschirner & Schindler, 2019) did not deal with
numbers at all, except for simple counting or addition. It dealt with modeling real-
world problems, specifically path problems and word puzzles, by undirected and
directed graphs. Mathematical games were also analyzed. Special emphasis was
placed on gradually transforming difficult problems into simpler ones and on con-
ducting mathematical proofs. In this volume, it is computed with numbers, which
students have surely been waiting for. The difficulty of the tasks in this *essential is*
slightly higher than in Volume I.

In Chap. 2 a real-world problem, namely the construction of different-sized
winner's podiums, motivates the need to efficiently compute sums of the form
$1 + 2 + \ldots + n$. After some example tasks and preliminary considerations, the
Gaussian summation formula is first conjectured, then proved and applied several
times. In Chap. 3 a recursion formula is derived to solve payment tasks. Similar to
other contexts in Volume I, this step by step reduces a difficult mathematical problem to simpler problems that can be solved. In Chaps. 4 and 5 prime numbers and
prime factorization are introduced. Various numbers are decomposed into prime
factors for practice. Students learn how to calculate the number of divisors of a
natural number from the prime factorization without having to determine them
explicitly. This also requires basic combinatorial considerations, which are also
worked out. A proof concludes Chap. 4. The penultimate problem chapter (Chap.
6) deals with problems with times and days of the week. Students quickly realize
that the hours of the day and the days of the week are periodic (with period 24 and
with period 7, respectively). Thus, calculating with remainders is motivated.
Modular arithmetic (modulo calculation) is introduced and beneficially applied. In

Chap. 7 students learn modulo calculation rules and apply them to sample tasks. The value of the calculation rules is also demonstrated on a subtask from Chap. 6 which can now be solved much easier and faster. At the end of Chap. 7 the divisibility rules for the numbers 3 and 9 are treated, and the modulo 9 remainder test", a relic from the time before the introduction of calculators, is addressed. Table. II.1 summarizes the mathematical techniques addressed in each chapter.

We have already pointed out in the first volume that concentrated work on this kind of tasks promotes and strengthens important and indispensable skills for further success in mathematics. This relates to developing one's own ideas, but also to "soft skills" such as patience, perseverance, and tenacity; see also Sects. 13.3 and 13.6 in (Käpnick, 2014). This provides experiences that should still have a positive impact on the understanding and learning of mathematics in higher grades and, looking very far into the future, may even be helpful for possible later studies of mathematics, computer science, or natural and engineering science. In addition, the mathematical methods and techniques learned also find ample application in primary and middle school (and occasionally even high school) mathematics competitions.

Less detailed than in Volume I, we only refer to the annual Mathematical Olympiads with class-specific tasks from grade 3 (Mathematik-Olympiaden e. V. 1996–2016; Mathematik-Olympiaden e. V. 2017–2018; Mathematik-Olympiaden e. V. 2013) and the Känguru competition (Noack et al., 2014). For a more detailed review of mathematics competitions, mathematical fiction, and a mathematics journal for students, the reader is referred to Volume I (Schindler-Tschirner & Schindler, 2019). In addition, the bibliography contains a number of other books with tasks and solutions from national and international mathematics competitions as well as task collections, but these are mostly aimed at older students.

With the two *essentials*, we would like to make a contribution to the promotion of giftedness among primary school students. In addition to the mathematical content, we would like to awaken the students' enjoyment of mathematics and encourage them to make mathematical discoveries.

1.2   Didactic Notes

Part II contains detailed sample solutions to the tasks from Part I with didactic hints and assistance for implementation in a working group. The solutions shown are designed in such a way that they are understandable and comprehensible even for non-mathematicians. The sample solutions are not directly intended for children. In addition, the mathematical goals of the respective chapters are explained, and outlooks are given where the mathematical techniques learned are applied in mathematics and computer science.

The abilities of the participating students should not be underestimated, but also not overestimated. It should definitely be explained to them (repeatedly) from the beginning that even very good students are by no means expected to be able to solve all tasks independently. This is very important, because a permanent over-challenge and/or (perceived) lack of success can lead to lasting frustrations, which are certainly not conducive to the attitude towards mathematics. That would be the opposite of what this *essential* aims to achieve. Therefore, participants should be carefully selected. In the mathematics club mentioned above, participants were nominated by the grade 3 and 4 class teachers.

Chapters 2 to 7 consist of many subtasks, the difficulty of which usually increases. Students with lower performance should preferably work on the easier subtasks. Some subtasks are very well suited for working in small groups of 2–3 students. This is sometimes indicated in the sample solutions. The teacher should give the students enough time to discover their own solutions and to pursue approaches that do not correspond to the sample solutions.

It is not easy, if not impossible, to develop assignments that are ideally suited to the needs of each mathematics SG or support course. It is at the discretion of the course instructor to omit subtasks or to add his own subtasks. In this way, he or she can influence the level of difficulty within certain limits and adapt it to the performance of the course participants. This applies to this *essential* in particular, since the individual chapters contain significantly more subtasks than in Volume I (Schindler-Tschirner & Schindler, 2019). This aspect is addressed several times in the sample solutions. In any case, the students' grasp and understanding of the solution strategies should be given priority over the goal of "creating" as many subtasks as possible. The individual chapters are likely to require more than one course meeting.

If the instructor works with task sheets, they should be read together; however, only those subtask(s) that are next up for completion should be read at a time. Presenting all subtasks at once could lead to rapid discouragement and resignation among the participants right at the beginning. A best practice is to have a high-performing student read the task aloud and, if necessary, clarify the task. Since younger students participate in the SG, this step is very important. Problems in understanding the tasks should not be underestimated.

Normally you should start with subtask (a). Each student should have a reasonable amount of time (depending on the ability level of the learning group) to think about the task alone (with help if necessary). Afterwards, the various ideas, approaches or perhaps even ready-made solutions are collected. Each student should regularly have the opportunity to present his or her approach or solution to the others. In this way, not only is one's own procedure reflected upon once again, but also such important competencies as a clear presentation of one's own considerations and mathematical argumentation are practiced; cf. also (Nolte, 2006), p. 94.

In the mathematics SG mentioned above, the fourth graders were on average significantly more efficient than the third graders. This was not because the third graders lacked the necessary prior mathematical knowledge. Rather, this was likely the result of greater intellectual maturity on the part of the older students. This may not be surprising for experienced teachers. In any case, the instructor should keep this effect in mind.

Embedding the tasks in a large, ongoing adventure story not only provides the narrative framework, but also gives the children a feeling of security. At the beginning of each new lesson, the teacher brings the children back into the fairy-tale, enchanted world of Clemens, so that fear of contact with the tasks does not even arise.

1.3   The Narrative Framework

Anna and Bernd are in the third grade. Their favorite subject is mathematics, and they are pretty good at it. They are eager to join the club of enthusiastic young mathematicians, or CoEYM for short. Unfortunately, according to the club's statutes, one may only join the CoEYM if one attends at least the fifth grade. There have been no exceptions so far.

However, Anna and Bernd were very persistent, so the club chairman Carl Friedrich gave them a chance. They are to help the sorcerer's apprentice Clemens, the CoEYM's club mascot, to complete 12 mathematical adventures,[1] so that he can win a set of useful magic paraphernalia that are indispensable for a magician.

Carl Friedrich exhorts Anna and Bernd to work together in solving the tasks. Carl Friedrich did not believe that they could make it into the CoEYM. However, he has since been surprised to find (at the end of Volume I) that the two have done very well and are well on their way to passing the entrance exam. However, they still have to prove themselves in the second half of the tasks, which are described in Chaps. 2, 3, 4, 5, 6 and 7 are to be solved.

Incidentally, Clemens has acquired the following magic paraphernalia in the first six mathematical adventures: a magic wand, a magic cloth, a magic ruby, a pinch of dragon ointment, a honeycomb with magic honey, and three witch hazels, which tell Clemens a solution hint when needed, but unfortunately can only be used once each.

---

[1] Six adventures each in Volume I and in this *essential* (Chaps. 2 to 7).

Part I

Tasks

There are six chapters with tasks. In the narrative context, these are the mathematical adventures of sorcerer's apprentice Clemens. New mathematical terms and techniques are introduced. The story and the tasks (and, of course, the instructor!) lead the children to the right way of solving the problems.

Each chapter ends with a section that describes the current situation from the point of perspective of Anna, Bernd, and Clemens. With a short summary of what the students have learned in this chapter, this section emerges from the narrative framework at the end. This description is not given in technical terms as in Table II.1, but in language suitable for students.

Summation Made Easy

A magicians' congress will soon be held in Right Angleton, where magicians from all over the world will perform their magic tricks in various disciplines. In each discipline the best magicians will be honored. The prizes and awards are presented on podiums made of magic stones. Unlike the Olympic Games, not only the three best magicians are honored, but many more. How many magicians are honored depends on the discipline.

Clemens is happy and proud to be helping organize this year's magicians' congress. He is to order magic stones for the winners' podiums from the Kadabra hardware store. Because magic stones are expensive, he should order the exact number. Figure 2.1 shows a winning podium for 5 magicians. This requires 6 magic stones. If Clemens does everything to the satisfaction of the wizard, he gets a magic rope. With a spell you can knot a magic rope in such a way that only very great wizards can untie these knots again.

(a) In the discipline "tricks with rabbits" the best 7 magicians are honored. How many rows of stones are needed for this winners' podium?
(b) How many stones are needed for a podium for 7 magicians?
(c) In the discipline "Tricks with double bottom" the best 23 magicians are honored. How many magic stones are needed for this winner's podium? First determine the number of rows of stones for this.
(d) In the discipline "Nothing is impossible" even 39 magicians are honored. How many magic stones are needed for this winner's podium?

© The Author(s), under exclusive license to Springer Fachmedien Wiesbaden GmbH, part of Springer Nature 2023
S. Schindler-Tschirner, W. Schindler, *Mathematical Stories II - Recursion, Divisibility and Proofs*, essentials,
https://doi.org/10.1007/978-3-658-38611-5_2

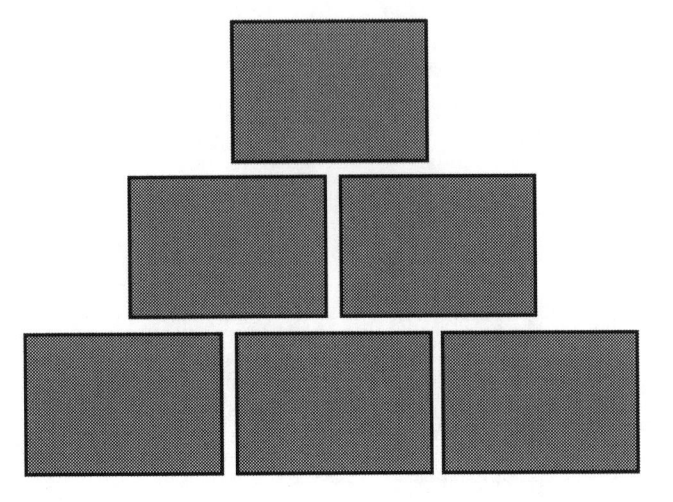

Fig. 2.1 Winners' podium for 5 magicians

Clemens realizes very quickly that he needs 20 rows of stones for this podium. In other words, he has to order 1 + 2 + ... + 20 magic stones, that's clear. But how many are there? Adding up that many numbers is really hard work! And he shouldn't miscalculate at all. After all, he wants to have a magic rope. Clemens already found subtask c) to be quite exhausting. So he puts this order on the back burner for now and looks in old magic books to see if it can't be done more easily.

In the book "Magic Stones – Ordering Made Easy" he finds the following yellowed page (see Fig. 2.2):

Calculation Rule The calculation rule inside a bracket is done first. For example, in the formula (3 · 4): 2, you first calculate 3 · 4 = 12, and then 12: 2 = 6.

Unfortunately, of all things, the most important formula is no longer readable, because mean trolls have smeared a thick blob of ink over the result.

(e) Check the formulas from Fig. 2.2 on subtasks (b) and (c).

(f) What do you think is written under the blob?

(g) How many magic bricks are needed to build the magic podium from task d), if your guess is correct?

(h) Clemens looked carefully at the formulas in the magic book and thinks that he has recognized a general law. He assumes that for all numbers the following formula is valid:

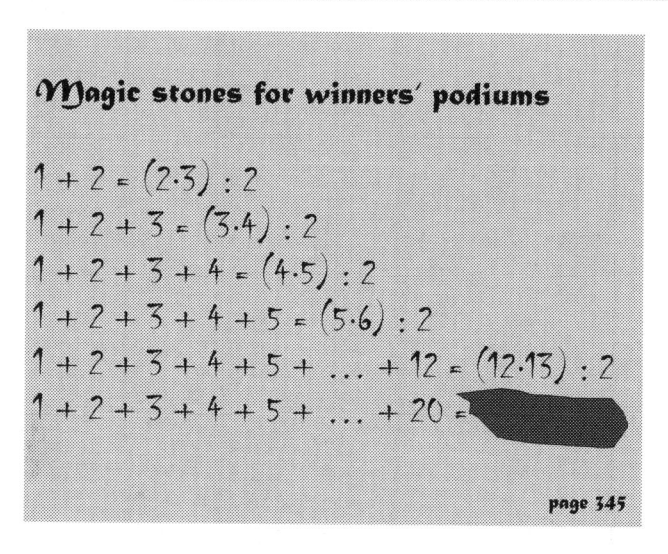

Fig. 2.2 Page 345 from "Magic Stones – Ordering Made Easy"

$$1 + 2 + \ldots + n = (n \cdot (n+1)) : 2 \quad \text{(Gaussian summation formula)} \quad (2.1)$$

What does that mean exactly? Mathematicians like to use such formulas in which the letter n stands for a number. For example, if we replace n by 3 or by 12, we get the second and the fifth formula from the magic book.

Apply the formula Eq. 2.1 to (b), (c), and (g).

(i) However, Clemens has already learned that in mathematics, guessing is a tricky thing. Can you help Clemens prove his conjecture?
(j) Use the Gaussian summation formula to calculate 1 + 2 + … + 30.
(k) Calculate 1 + 2 + … + 55.
(l) Calculate 1 + 2 + … + 100.
(m) Calculate 3 + 4 + … + 39. But be careful! This sum starts with 3. (Hint: Add "1 + 2 +" to the sum on the left and subtract 3 at the end).

Anna, Bernd, Clemens and the Students
Clemens is happy because he now also has a magic rope.

Anna and Bernd are pleased that they finally have calculated with numbers. Bernd says: "The formula $1 + 2 + \ldots + n = (n \cdot (n + 1)) : 2$ is super. You can do a lot with it. I have heard that the great mathematician Carl Friedrich Gauss[1] discovered this formula on his own when he was our age." And Anna adds, "I wouldn't have thought you could discover new mathematical formulas at our age. Maybe we'll discover a new formula someday, too. That would be great!"

The club chairman of the CoEYM is also called Carl Friedrich. I wonder if that means something.

What I have Learned in this Chapter

- You can easily calculate $1 + 2 + \ldots + n$ using the Gaussian summation formula.
- I proved and applied the Gaussian summation formula.

---

[1] Carl Friedrich Gauss (1777–1855) was an outstanding German mathematician (cf. Chap. 8)

Payment Problems at the Kiosk

3

Clemens wants to buy some sweets from the kiosk of the good-natured troll Eberhard with his pocket money. "But I am paid with magic euros (M€) and magic cents (MC)," Eberhard instructs him. Clemens already wants to leave the kiosk disappointed, but Eberhard holds him back and points to a change machine to the left of the counter. "Furthermore, I'll give you a few tasks. If you can solve them, I'll give you another pack of magic soda," says Eberhard.

(a) A tasty raspberry candy costs 8 MC. Give all the ways Clemens can pay this amount in 1 and 2 magic cent coins. The order in which Clemens places the coins on the counter does not matter.

(b) A small bag of gummy moose costs 13 MC and a chocolate bar costs 21 MC. Again, give all the ways to pay these amounts in 1 and 2 magic cent coins.

(c) Do you recognize a law that allows you to calculate the number of ways to pay without listing all the possibilities?

Notation An underlined number denotes a coin of that value, e.g., ($\underline{2}$-MC) denotes a coin of 2 magic cents. It denotes $A(n|\underline{1},\underline{2})$ the number of ways to pay an amount of n MC in ($\underline{1}$-MC) and ($\underline{2}$-MC) coins.

Example $A(8|\underline{1},\underline{2}) = 5$

Now good advice is expensive, because Clemens has no idea how to solve subtask (c). Then he remembers the three witch hazels he won in the last adventure in Volume I (Schindler-Tschirner & Schindler, 2019). He takes a magic nut out of his pocket and hurls it with all his might onto the ground. After a loud roar and colorful smoke, a dark voice speaks, "Distinguish between even and odd amounts. This will bring you closer to the solution".

From (a) and (b) we know that $A(8|\underline{1},\underline{2}) = 5$, $A(13|\underline{1},\underline{2}) = 7$ and $A(21|\underline{1},\underline{2}) = 11$. In subtask (c), we are looking for calculation formulas for $A(n|\underline{1},\underline{2})$, where we have to distinguish between even and odd n.

(d) Apply the formulas you found to subtasks (a) and (b).
(e) Apply the formulas to the chocolate (72 MC) and the big bag of gummy moose (53 MC).
(f) In how many ways can Clemens pay for the chocolate bar for 21 MC if he is also allowed to use ($\underline{5}$-MC) coins?

Clemens understands payment tasks with ($\underline{1}$-MC) and ($\underline{2}$-MC) coins, but subtask (f) is much more complicated. Clemens sacrifices another witch hazel, and the dark voice says to him, "Take advantage of what you already know."

Clemens doesn't know what to do with this hint and is sad that he used the witch hazel for nothing. Then Eberhard says to him, "Think about how many possibilities there are if you use exactly two ($\underline{5}$-MC) coins. What about if you use exactly three ($\underline{5}$-MC) coins?"

(g) Solve the problems that Eberhard has just given.

Clemens has understood Eberhard's hint. He writes the following formula on a sheet of paper. Here $A(21|\underline{1},\underline{2},\underline{5})$ denotes the number of ways to pay 21 MC with ($\underline{1}$-MC)-, ($\underline{2}$-MC)- and ($\underline{5}$-MC)-coins.

$$A(21|\underline{1},\underline{2},\underline{5}) = A(21|\underline{1},\underline{2}) + A(16|\underline{1},\underline{2})$$
$$+ A(11|\underline{1},\underline{2}) + A(6|\underline{1},\underline{2}) + A(1|\underline{1},\underline{2}) \qquad (3.1)$$

"Very good, Clemens!" praises Eberhard.

Eberhard explains This is an example of a recursion formula. The unknown value $A(21|\underline{1},\underline{2},\underline{5})$ is expressed as the sum of $A(21|\underline{1},\underline{2})$, $A(16|\underline{1},\underline{2})$, $A(11|\underline{1},\underline{2})$, $A(6|\underline{1},\underline{2})$, and $A(1|\underline{1},\underline{2})$. The summands look similar to $A(21|\underline{1},\underline{2},\underline{5})$, but they are easier to calculate, because you don't have to consider three different coins anymore, but only two. You already know how many payment options there are when you can only pay with ($\underline{1}$-MC) and ($\underline{2}$-MC) coins. Sometimes you have to do several such steps.

(h) Explain the formula Eq. 3.1 and calculate $A(21\,|\underline{1},\underline{2},\underline{5})$.
(i) Calculate $A(19\,|\underline{1},\underline{2},\underline{5})$.
(j) Can you calculate $A(21|\underline{1},\underline{2},\underline{5},\underline{10})$? Now ($\underline{10}$-MC) coins may also be used. Use Clemens' idea for this.
(k) Think up a payment task yourself and solve it.

Anna, Bernd, Clemens and the Students
Clemens finally has a packet of magic soda with which he can turn ordinary tap water into any drink of his choice.

Anna says, "A recursion formula like that is a great thing." And Bernd proudly adds, "And we derived it ourselves!" In addition, Anna and Bernd are amazed (and also a little envious) at how cheap candy is in Right Angleton.

What I have Learned in this Chapter
- I have learned a recursion formula.
- With a recursion formula you can reduce difficult problems to simpler problems and solve them.

The First Encounter with Dwarf Dividus

4

In Dwarf Village, a neighboring village of Right Angleton, dwarves live. Clemens meets the dwarf Dividus there. He likes mathematical puzzles, but most of all he likes to divide numbers. Figure 4.1 shows his last work.

And Dividus is generous. "I'll give you a stealth cap if you can solve some interesting problems." "A stealth cap would be great," Clemens says with anticipation. "You'll need to know a few things about it, though," Dividus replies. "I'll give you a few hints."

Dividus explains The numbers 1, 2, 3, … are called natural numbers. A natural number m is called a divisor of n if n is divisible by m without a remainder.

Example 4 is a divisor of 12, and 9 is a divisor of 18, but 5 is not a divisor of 9.

(a) Determine for all natural numbers from 1 to 30 the set of their divisors.

Example: The number 10 has exactly four divisors, namely 1, 2, 5 and 10.

(b) Which of these numbers have the fewest divisors, and which have the most? Are there numbers that have exactly two divisors?

S. Schindler-Tschirner, W. Schindler, *Mathematical Stories II - Recursion, Divisibility and Proofs*, essentials, https://doi.org/10.1007/978-3-658-38611-5_4

Fig. 4.1 Prime factorization
of the number 12

$$12 = 2 \cdot 2 \cdot 3$$

Dividus explains   Natural numbers that are divisible only by 1 and themselves
are called prime numbers. But remember: The number 1 is not a prime number!

(c) Give 5 prime numbers.

(d) Which of the following numbers are prime: 7, 14, 41, 51, 72, 83, 100?

(e) Determine all prime numbers that are smaller than 30. Use the results from a)
     and b).

(f) Represent the natural numbers from 2 to 15 as a product of prime numbers.

Example: $10 = 2 \cdot 5$, $11 = 11$.

Dividus explains   You can represent any natural number n greater than 1 as the
product of primes. This is called the prime factorization of n. By the way, the
prime factorization is unique, if you ignore the order of the prime factors.

   Dividus states, "You have just calculated the prime factorizations of the numbers 2 through 15, Clemens!"

(g) Calculate the prime factorization of the natural numbers between 16 and 30.

Dividus explains   A natural number n is called a square number if there is a
natural number m for which $m \cdot m = n$.

Example   25 is a square number, because $25 = 5 \cdot 5$, but 10 is not a square
number.

(h) Which of the natural numbers between 1 and 30 have an odd number of divi-
     sors?

(i) Do you have a guess which natural numbers between 1 and 200 have an odd
     number of divisors?

(j) Try to prove your conjecture.

Anna, Bernd, Clemens and the Students

Bernd says, "Another proof at the end." "Proving is not easy at all, but when you have found a proof, it makes you proud," Anna says.

What I have Learned in this Chapter

- I know what prime numbers are.
- I have decomposed numbers into their prime factors.
- I have seen and understood another proof.

Dwarf Minimus Is Not Nice at All

5

After Clemens wins the stealth cap from Dwarf Dividus, Dwarf Minimus (the smallest dwarf in all of Dwarf Village) also offers a prize, namely a green emerald with incredible magic powers. However, Dwarf Minimus is not at all friendly. He is quite sure that no one can solve his math puzzles, and certainly not a child. Therefore, he persuades Clemens to make a risky bet: if Clemens can solve Minimus' problems, he will get the emerald; otherwise, he will have to give Minimus his stealth cap, which he won only in the last adventure. "I'd like to see the tasks first," says Clemens, but Minimus sneeringly replies, "Then anyone could bet! If you're afraid or just don't know anything about math, then let's forget the bet. By the way, the tasks have to do with divisors." Clemens has just learned a lot about this from the dwarf Dividus. Clemens gets cocky and accepts the bet.

"All right, Clemens. Then tell me how many divisors the number 42 has." Clemens mumbles quietly to himself, "1 is a divisor of 42, 2 is a divisor of 42, 3 is a divisor of 42, 4 is not a divisor of 42, …." "Will it be soon, Clemens? How long am I supposed to wait?" "After all, I have to check all the numbers between 1 and 42 to see if they divide 42 or not. That takes time." "But you don't have that time! How will it be when I ask you how many divisors the numbers 125 or 168 have? You'll have to do it in 3 minutes or less. Otherwise, I'll have won the bet. Are you going to give up right now?" "No, Minimus, no. I need my magic hat after all," Clemens pleads. "All right, Clemens, I'll give you one more chance. You have until tomorrow to figure out how to solve my problems. Maybe you'll get an idea over-night," he adds with a sneer.

Quite despondent and deeply saddened, Clemens sets off for Dividus. Arriving at Dividus, he tells him about his bet. "I think I can help you," Dividus says. "But

© The Author(s), under exclusive license to Springer Fachmedien
Wiesbaden GmbH, part of Springer Nature 2023
S. Schindler-Tschirner, W. Schindler, *Mathematical Stories II - Recursion,
Divisibility and Proofs*, essentials,
https://doi.org/10.1007/978-3-658-38611-5_5

you must promise me not to bet again." "I'll do that," Clemens says meekly, "just bail me out." "It's not that simple, though. I can't tell you the solution method. That would violate the dwarves' code of honor, because after all, you were betting against a dwarf. And besides, if you want to get the green emerald, you'll have to make an effort yourself. But I can give you a few tips. After all, Minimus talked you into this bet."

"Decompose the numbers 42 into their prime factors, Clemens." Clemens does the math on the small slate Dividus always has with him:

$$42 = 2 \cdot 21 = 2 \cdot 3 \cdot 7$$

"Very good! As you know, $2 \cdot 3 \cdot 7$ is the prime factorization of 42. And for practice, a few more tasks."

(a) Decompose 63 into prime numbers.
(b) Decompose 125 into prime numbers.

Dividus explains For each natural number n, the notation $n^1 = n$, $n^2 = n \cdot n$, $n^3 = n \cdot n \cdot n$, …. This allows you to write down the prime factorization more clearly. This is called powers. The large number is called the base, and the small superscript number is the exponent. Also, $n^0 = 1$ for all natural numbers n.

Example $2^0 = 1$, $2^1 = 2$, $2^2 = 2 \cdot 2 = 4$, $2^3 = 2 \cdot 2 \cdot 2 = 8$,… and $5^0 = 1$, $12^1 = 12$,
$23^2 = 23 \cdot 23$.

It is "23^2" a power of 23, where "23" is the base and "2" is the exponent.

(c) Use the power notation for the prime factorizations of 63 and 125.

Dividus gives another hint: "Clemens, decompose the number 12 into its prime factors, and write all the divisors of 12 on the board. Do you notice anything?" Clemens writes

$$12 = 2^2 \cdot 3, \quad \text{divisors of } 12 = \{1, 2, 3, 4, 6, 12\} \tag{5.1}$$

"Now decompose all the divisors of 12 that are greater than 1, even into prime factors."

$$\text{divisors of } 12 = \left\{1, 2, 3, 2^2, 2 \cdot 3, 2^2 \cdot 3\right\} \tag{5.2}$$

Clemens thinks hard, but he still does not recognize any regularity.

(d) Can you help him? Decompose 20 into its prime factors. Write down all divisors of 20 and decompose them (except 1) into prime factors. How many divisors are there?

(e) Decompose 35 into its prime factors. Write down all divisors of 35 and decompose them (except 1) into prime factors. How many divisors are there?

(f) The fashion-conscious mouse Ron Rodent owns three shirts, namely a blue, a yellow and a red shirt. Ron also has a pair of striped pants and a pair of dotted pants.

How many different combinations of shirt and pants are there?

(g) Furthermore, Ron Rodent owns four pairs of socks, namely a black pair, a white pair, a black and white checkered pair, and a purple pair.

In how many different ways can Ron Rodent dress, i.e., choose a shirt, pants, and socks?

"What does that have to do with the dividers?" asks Clemens, annoyed. "Even though I'm sure that's totally interesting, I don't have time for that right now, Dividus". "Have faith in me. That's all I can help you with because of the dwarves' code of honor," Dividus replies.

Then Clemens remembers that he still has one last witch hazel. He throws it to the ground, and the already familiar dark voice speaks: "Even the seemingly superfluous can sometimes be useful. Represent all divisors of 12 as products of powers of 2 and 3, even if 2^0 or 3^0 occur."

Still a little shaky, Clemens writes

$$\text{divisors of } 12 = \left\{2^0 \cdot 3^0, 2^1 \cdot 3^0, 2^0 \cdot 3^1, 2^2 \cdot 3^0, 2^1 \cdot 3^1, 2^2 \cdot 3^1\right\} \tag{5.3}$$

(h) Back to the divisors: do you recognize a rule now? How are the divisors formed? Can you *work out* how many divisors 12 has?

(i) Try to calculate the number of divisors of 55 from the decomposition into prime factors without determining the divisors themselves.

After a restless night, Clemens goes to Minimus. Confident of victory and with a gleeful grin, Minimus gives Clemens a sheet with the following six tasks:

(j) How many divisors does the number 100 have?
(k) How many divisors does the number 99 have?
(l) How many divisors does the number 128 have?
(m) How many divisors does the number 168 have?
(n) How many divisors does the number 525 have? Hint: $525 = 3^1 \cdot 5^2 \cdot 7^1$.
(o) How many divisors does the number 529 have? Hint: $529 = 23^2$.

Now it's on! Clemens has only roughly understood the hints of Dividus and the dark voice. You have to help him now!

Anna, Bernd, Clemens and the Students
That was a very long, exhausting adventure. Clemens is quite exhausted, as are Anna and Bernd.

Anna and Bernd are amazed at how many new mathematical techniques they have already learned. "Will there be more new math in the last two adventures?" wonders Bernd.

What I have Learned in this Chapter

• I have again decomposed numbers into prime factors.
• I now know how to calculate the number of divisors from the prime factorization.

Dwarf Modulus Intervenes

6

Every Friday evening, Clemens watches the extremely popular quiz show "Time, Day and Year", which is broadcast on the Right Angleton station "Quiz-TV". There, the candidates have to answer questions as quickly as possible, such as "What day of the week is in 235 days?" or "What time is it in 43 hours?". Greyhound Velox is the station's champion. Whoever beats him in a quiz duel wins one of the coveted magic clocks. The quiz duel is won by whoever scores 5 points first. The first person to answer a question correctly receives one point. However, each contestant is only allowed to give one answer to each question, so they can't just try all the days of the week or all the times.

Clemens is fascinated by the show and especially by the prospect of a magic clock. With such a clock, he hopes, he could generate two Sundays in 1 week and thus get pocket money twice. However, Velox is insanely fast. The other day, it took him all of 10 seconds to solve (correctly!) the question of which day of the week is in 235 days. "Unbelievable," thinks Clemens, "I would probably need at least 5 minutes for that. But with good math, I'm sure it can be done much faster, and maybe I can even beat Velox." Only: Even after a long time of hard thinking, Clemens still doesn't have a good idea.

So once again he goes to the dwarf Dividus and tells him his problem. Dividus doesn't know what to do either. Fortunately, his cousin, the dwarf Modulus, is visiting Dividus for a few days. He overheard the conversation between Clemens and Dividus and finally says: "I know what can be done. Have you heard anything about the modulo calculation, Clemens?" "No, not yet."

"It's not that hard," Modulus reassures him. "Let's start with a simple day-of-week problem. Today is Tuesday, Clemens. What day of the week is 16 days from now?" Clemens begins to quietly list the days of the week, "1st day: Wednesday, 2nd day: Thursday, 3rd day: Friday, …, 8th day: Wednesday, 9th day: Thursday." "Wait a minute", Modulus interrupts him, "We've had Thursday before. Do you remember on which day?" Clemens thinks for a moment, "Yes, on the second day." "Do you notice anything?" Clemens thinks, and suddenly it's clear to him: "Exactly one week has passed between the second and ninth day. That's why it's the same day of the week on both days." "Very good, Clemens, pursue this thought further." "Oh yes, after 16 days, another week has passed, and it's Thursday again." "Right!" says Modulus appreciatively, "you've discovered the principle. For practice, one more task: what day of the week is in 70 days?" Clemens thinks for a moment, "After 70 days, exactly 10 weeks have passed, so it's Tuesday again, like today."

(a) Divide with remainder:

$$16:7=, \qquad 9:7=, \qquad 2:7=, \qquad 70:7=$$

"Do you notice anything, Clemens? The first three tasks have different solutions, of course, but the numbers 16, 9, and 2 still have one thing in common: If you divide them by 7, they have the same remainder.

In our weekday tasks, everything depends on what remainder a number gives when you divide it by 7 (remainder modulo 7). But in other tasks, numbers other than 7 can be important."

(b) Divide with remainder:

$$16:5=, \qquad 11:5=, \qquad 9:5=$$

"Now 16 and 11 have the same remainder when you divide them by 5 (remainder of 5). But be careful: the numbers 16 and 9 have the same remainder of 7, but not the same remainder of 5."

Modulus writes on his slate (see Fig. 6.1):

Further Examples

$5 \equiv 2 \bmod 3$, since $5:3 = 1$ remainder 2

$12 \equiv 2 \bmod 10$, since …

$16 \equiv 9 \equiv 2 \bmod 7$, since …

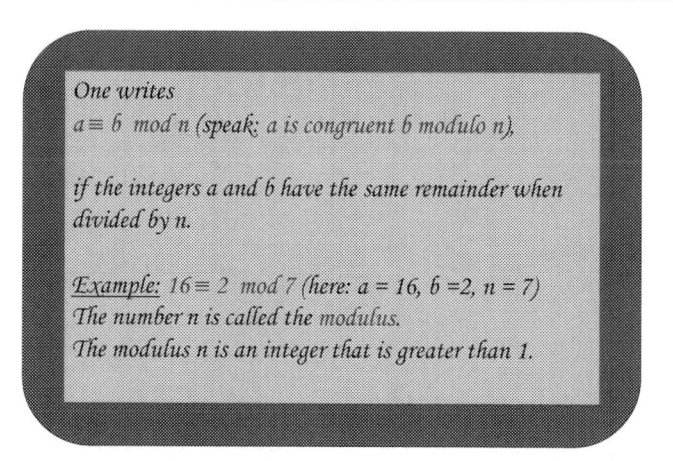

One writes

$a \equiv b \mod n$ *(speak: a is congruent b modulo n),*

if the integers a and b have the same remainder when divided by n.

Example: $16 \equiv 2 \mod 7$ (here: a = 16, b = 2, n = 7)
The number n is called the modulus.
The modulus n is an integer that is greater than 1.

Fig. 6.1   A look at Modulus's slate

Modulus explains   The numbers 0, 1, 2, … are called nonnegative integers. Here are a few simple tasks to help you become familiar with modulo .

(c) Determine the smallest non-negative number (0, 1, 2, …) for which the congruence is correct. Enter the value to the right of the congruence sign $\equiv$.

$$22 \equiv \quad \mod 10, \qquad 17 \equiv \quad \mod 2, \qquad 22 \equiv \quad \mod 15,$$
$$52 \equiv \quad \mod 25, \qquad 17 \equiv \quad \mod 7, \qquad 22 \equiv \quad \mod 28.$$

(d) "Clemens, it's just 18 o'clock (6 pm), time for dinner. What time is it in 26 hours?" Clemens thinks for a moment and mutters, "There are 24 hours in a day …." What do you think Clemens meant by this? Use modulo calculations to solve the following problems.

Note: Subtasks (d), (e), and (f) consider the 24-hour clock. This means that the hours of day are denoted by 0 o'clock (0 am), 1 o'clock (1 am), …, 12 o'clock (12 am), 13 o'clock (1 pm), …, 23 o'clock (11 pm).

(e) Now it is 10 o'clock (10 am). What time is it in 52 h?

(f) Now it is 23 o'clock (11 pm). What time is it in 27 h?

Here are some simple exercises for the modulo calculation.

(g) Find the smallest non-negative number (0, 1, 2, …) for which the congruence
     is correct. Enter the value to the right of the congruence sign ≡.

$$29 \equiv \quad \text{mod } 24, \qquad 241 \equiv \quad \text{mod } 24, \qquad 59 \equiv \quad \text{mod } 24.$$

(h) "January 1, 2019 is a Tuesday. What day is January 1, 2020, Clemens?" Can
     you help Clemens solve this problem?

Now Clemens feels well enough prepared to challenge Velox in a quiz duel. At the
beginning, Clemens was still quite nervous, and Velox was able to quickly pull
away to 2:0. After eight questions, the score is 4:4, and the next question must decide the outcome.

(i) Quizmaster Winston Wise asks: "January 1, 2019 is a Tuesday. What day is
     January 1, 2023?" "Saturday," exclaims Velox hastily. "That answer is wrong!"
     says Winston Wise. "If Clemens knows the right answer now, he's won." Help
     Clemens win the magic clock.

Anna, Bernd, Clemens and the Students
Clemens is relieved that he did manage to win the magic clock at the last moment.
"I've never heard of modulo in school," Anna says, and Bernd means, "Modulo is
really cool."

What I have Learned in this Chapter

- I can calculate what day of the week it will be in exactly 1 year.
- I have learned about modulo calculation.

Even More Calculations with Remainders

The guessing quiz duel from the last adventure turned out well for Clemens. He thanks Dwarf Modulus for his help, without which he certainly wouldn't have won the coveted magic clock. Clemens says, "The modulo calculation was my salvation. Can you actually use it for other things?" "Oh yes, there are even a lot of applications for the modulo calculation," replies Dwarf Modulus, "if you're interested, I'll show you more of modulo. We'll start with a useful calculation rule. If you can solve a few problems, I'll give you a math book on the modulo calculation. It will definitely come in handy for future adventures."

Modulus explains   Modulo calculation rule 1 (addition):
$$a \equiv a' \bmod n \quad \text{and} \quad b \equiv b' \bmod n \quad \text{imply} \quad a + b \equiv a' + b' \bmod n.$$

Example   It is $22 \equiv 2 \bmod 10$ and $19 \equiv 9 \bmod 10$. From calculation rule 1, it follows $22 + 19 \equiv 2 + 9 \equiv 11 \equiv 1 \bmod 10$.

Modulus explains   This calculation rule also applies to sums with multiple summands:

$$23 + 87 + 3 + 10 \equiv 1 + 1 + 1 + 0 \equiv 3 \equiv 1 \bmod 2$$

So one can replace the summands by their remainders. This simplifies the necessary calculations quite a bit because you no longer have to add large numbers.

(a) Determine the smallest nonnegative number for which the congruence is correct. Calculate cleverly!

$$22 + 17 \equiv \quad \mod 10, \quad 100 + 17 \equiv \quad \mod 10, \quad 31 + 17 \equiv \quad \mod 3,$$
$$7 + 2 \equiv \quad \mod 4, \qquad 12 + 2 + 3 \equiv \quad \mod 2.$$

Use arithmetic rule 1 to solve subtask (i) from the last mathematical adventure more easily:

(b) January 1, 2019 is a Tuesday. What day is January 1, 2023?

"Modulo is really cool," Clemens enthuses. "But it gets even better," explains dwarf Modulus: "What's true for addition is also true for multiplication."

Modulus explains   Modulo calculation rule 2 (multiplication):

$$a \equiv a' \bmod n \quad \text{and} \quad b \equiv b' \bmod n \quad \text{imply} \quad a \cdot b \equiv a' \cdot b' \bmod n$$

(c) Find the smallest non-negative number for which the congruence is correct. Calculate cleverly!

$$2 \cdot 22 \equiv \quad \mod 7, \qquad 10 \cdot 17 \equiv \quad \mod 3, \qquad 31 \cdot 17 \equiv \quad \mod 31.$$

Here the calculation advantage is even greater, because multiplying large numbers is more time-consuming than adding them.

(d) Determine the smallest non-negative number for which the congruence is correct. Calculate cleverly!

$$10 \equiv \quad \mod 3, \qquad 100 \equiv \quad \mod 3, \qquad 1000 \equiv \quad \mod 3,$$
$$10 \equiv \quad \mod 9, \qquad 100 \equiv \quad \mod 9, \qquad 1000 \equiv \quad \mod 9.$$

(e) Determine the smallest non-negative number for which the congruence is correct. Calculate cleverly! Use subtask (d) and calculation rule 2.

$$3000 \equiv \quad \text{mod } 9, \qquad 200 \equiv \quad \text{mod } 9, \qquad 40 \equiv \quad \text{mod } 9.$$

(f) Determine the remainder of the number 3246 when divided by 9. Calculate cleverly.

Hint: Represent 3246 in thousands, hundreds, tens and ones and use the results from subtask (e).

(g) Is the number 3564 divisible by 9?

"Do you notice anything, Clemens?" asks Dwarf Modulus. Clemens thinks for a bit and exclaims, "That's totally interesting! The two numbers have the same remainder modulo 9 as the sum of their digits."

Modulus explains   The sum of the digits of a number is called the digit sum of that number.

Example   The digit sum of 5234 is $5 + 2 + 3 + 4 = 14$.
     "By the way, what you observed in the two examples is true in general," dwarf Modulus remarks a bit proudly.

Modulus explains   The remainder modulo 9 of a number is equal to the remainder modulo 9 of its digit sum. The same is true for the remainder modulo 3, but it's not usually true for other remainders.
     "By the way, you can prove this totally interesting statement with modulo calculation," Modulus says.

(h) Can the following results be correct? Check this without actually doing the multiplication results. Consider the remainder modulo 9 instead.

Fig. 7.1   Emblem of the CoEYM

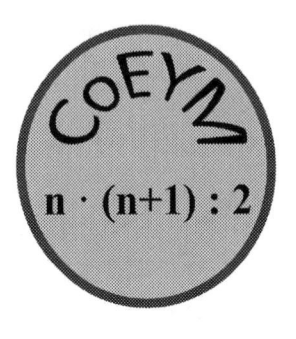

$$34 \cdot 54 = 1736, \qquad 27 \cdot 44 = 1178, \qquad 24 \cdot 19 = 456, \qquad 37 \cdot 41 = 1508.$$

Anna, Bernd, Clemens and the Students

Clemens is very happy. He has successfully passed all the mathematical adventures and won many useful magic utensils and, last but not least, a math book on modulo calculation. The sorcerer's apprentice Clemens has not yet become a real sorcerer, but he is already a respectable journeyman sorcerer.[1]

The club chairman Carl Friedrich welcomes Anna and Bernd in a friendly manner and hands them their membership cards: "Welcome to the CoEYM! I congratulate you very much, Anna and Bernd. That was super! I have to admit that I didn't expect this at first." Anna and Bernd beam: "We learned a lot of math and worked well together. It was really fun!" Anna says, "I was surprised to learn that mathematics isn't just calculating and that you have to be so creative." Bernd states, "We proved many statements. We didn't know proofs at all before."

What I have Learned in this Chapter

- I have calculated with remainders again.
- I now also know modulo calculation rules.
- With these calculation rules, modulo calculations become much easier.

It is up to the course instructor to decide whether to issue membership cards for the CoEYM to the participants of the study group at the end (Fig. 7.1).

---

[1] A journeyman magician is the first step an apprentice magician must climb on the way to becoming a magician.

Part II

Sample Solutions

This part contains detailed sample solutions to the tasks from Part I. The target group are leaders of study groups for gifted school students, teachers, and parents (but not the students). As a rule, this makes hardly any difference; only in some places is differentiation made. In order to avoid awkward wording, only the "instructor" is normally addressed in the following. Table II.1 shows the most important mathematical techniques used in the individual chapters.

Table II.1 Overview: Mathematical contents of the task chapters

Chapter	Mathematical techniques	Outlook
Chapter 2	Gaussian summation formula (proof and applications)	Mathematics in upper school level and mathematics competitions, historical: Carl Friedrich Gauss
Chapter 3	Real world problem (payment problem), stepwise reduction to smaller problems, recursion formula	Mathematics (Fibonacci sequence) and computer science (recursive functions)
Chapter 4	Prime factorization, divisors, mathematical proof	School mathematics in the lower school level
Chapter 5	(Continuation of Chap. 4) prime factorization, number of divisors, combinatorics	Mathematics competitions
Chapter 6	Modular arithmetic (modulo calculation) with applications (calculation of times of day and days of the week)	
Chapter 7	Modular arithmetic (modulo calculation rules), divisibility rules for 3 and 9, mathematical proof, modulo 9 remainder test	Simple proof of an upper school problem of the mathematics olympiad, cryptography

The sample solutions, the mathematical goals of the individual chapters are explained, and outlooks are given on where the mathematical techniques learned are still being used. It can give the children additional motivation and self-confidence when they learn that very advanced tasks can be solved with the techniques learned (see also the preface by (Amann 2017)).

At the end of each task chapter, you will find a summary "What I have learned in this chapter". This is a counterpart to Table II.1, but in a language suitable for students. The teacher can work out the learning progress together with the students. This can be done, for example, at the following course meeting to recapitulate the last chapter.

Sample Solution for Chapter 2

8

Unlike Volume I (Schindler-Tschirner & Schindler, 2019), in this volume we "calculate". Thus, the children are on familiar ground. The instructor should use this to encourage and motivate especially those children who had difficulties with the considerations and reasoning from the first volume that was still unfamiliar to them. The first two subtasks are relatively easy. The instructor should pay attention that they are presented by lower-performing students, if possible.

(a) There is room for one magician on the top step. So there must be places for six more wizards. For every two wizards you need a new row of stones. In total you need $1 + (6 : 2) = 1 + 3 = 4$ rows of stones.

(b) The top row consists of a single wizard stone, and each additional row requires one more stone than the row above. So this podium requires a total of $1 + 2 + 3 + 4 = 10$ magic stones.

(c) The considerations are completely analogous to the subtasks a) and b). You need $1 + (22 : 2) = 1 + 11 = 12$ rows of stones and $1 + 2 + \ldots + 12 = 78$ magic stones.

(d) You need $1 + (38 : 2) = 1 + 19 = 20$ rows of stones. So Clemens has to calculate the sum $1 + 2 + \ldots + 20$. Adding the numbers is too tedious for him. Therefore, he looks for a more efficient method.

(e) Applying the formulas from Fig. 2.2 to subtasks b) and c) yields the already known solutions: $(4 \cdot 5) : 2 = 20 : 2 = 10$ and $(12 \cdot 13) : 2 = 156 : 2 = 78$.

(f) **Observation** In all formulas, the right-hand sides are of the following type: ((greatest summand) $\cdot$ (greatest summand $+1$)) : 2.

S. Schindler-Tschirner, W. Schindler, *Mathematical Stories II - Recursion, Divisibility and Proofs*, essentials, https://doi.org/10.1007/978-3-658-38611-5_8

So it can be assumed that under the blob "$(20 \cdot 21) : 2$" is written.

(g) If the guess is correct, $(20 \cdot 21) : 2 = 420 : 2 = 210$ magic bricks are needed. This subtask should cause little trouble for the children.

(h) Subtask h) is supposed to reduce "fear of contact" with variables. Formula Eq. 2.1 confirms the results from b), c) and g) by substituting the numbers 4, 12 and 20 for n.

(i) **Didactic Suggestion** It is not expected that primary school students will be able to prove the Gaussian summation formula on their own. Therefore, the instructor should work out the proof together with the students.

For students, the proof of formula Eq. 2.1 with letters may be too abstract at first. Therefore, the proof idea should first be illustrated with a concrete numerical example, e.g. for n = 4. To do this, we write the sum $1 + 2 + 3 + 4$ twice, rearrange some summands and combine two summands at a time in one bracket:

$$1+2+3+4+1+2+3+4 = 1+4+2+3+3+2+4+1$$
$$= \left(1+4\right)+\left(2+3\right)+\left(3+2\right)+\left(4+1\right).$$

If we look at the second summands in the brackets, we see the numbers from 1 to 4 in reverse order. Each of the four brackets gives the value 5, and all brackets together give $5 + 5 + 5 + 5 = 4 \cdot 5$. However, we have calculated the double of $1 + 2 + 3 + 4$. So we have to divide the result by 2 and finally get

$$1+2+3+4 = \left(4 \cdot 5\right) : 2 = 10.$$

Figure 8.1 shows a familiar geometric illustration of rearranging and combining two summands at a time for n = 4. (Turning one "staircase" around, the "steps" complete a rectangle. The horizontal rows correspond to the bracket expressions).

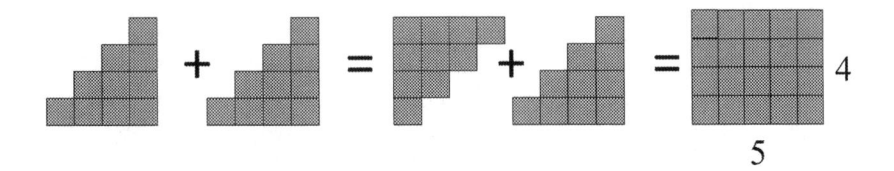

Fig. 8.1 Illustrative proof of the Gaussian summation formula for n = 4

Depending on how well the students have understood the proof for the special case n = 4, the instructor can work with the students to prove the formula again for n = 7, before proving the formula for any n.

Proof   As in the special case n = 4, we write the left side of Eq. 2.1 twice, rearrange the summands and combine two summands at a time in one bracket. The right summands in the brackets are n, n − 1, ..., 1.

$$1+2+...+n+1+2+...+n = 1+n+2+(n-1)+...+(n-1)+1+n+1$$
$$= (1+n)+(2+(n-1))+...+((n-1)+1)+(n+1)$$
$$= (n+1)+...+(n+1) \tag{8.1}$$

Each bracket has the value n + 1, and there are exactly n brackets. (For n = 4, the value of each bracket is 4 + 1 = 5.) Adding all the brackets together gives n · (n + 1). However, we added up the sum 1 + ... + n twice. Thus

$$1+2+...+n = (n\cdot(n+1)) : 2,$$

what had to be proved.

Didactic Suggestion   Depending on the ability level of the course, the general proof can be omitted. However, students should at least be able to apply the Gaussian summation formula.

(j) 1 + 2 + ... + 30 = (30 · 31) : 2 = 930 : 2 = 465
(k) 1 + 2 + ... + 55 = (55 · 56) : 2 = 3080 : 2 = 1540
(l) 1 + 2 + ... + 100 = (100 · 101) : 2 = 10100 : 2 = 5050
(m) This subtask is a bit more difficult, since one cannot apply Eq. 2.1 directly. Therefore, we use a trick that is common in mathematics: We add something to subtract it right away. This does not change the sum, but we can then apply our formula.

$$3+4+...+39 = 1+2+3+4+...+39-1-2 = (39\cdot40) :$$
$$2-3 = 1560 : 2-3 = 780-3 = 777.$$

Mathematical Goals and Outlook

The goal of Chap. 2 is to prove the Gaussian summation formula and to apply and practice it on examples. Even students who cannot contribute to the proof of the formula should have a sense of achievement by calculating simple sums and applying the formula correctly.

The formula $1 + 2 + \ldots + n = (n \cdot (n + 1)) : 2$ goes back to one of the greatest German mathematicians, Carl Friedrich Gauss (1777–1855). Gauss was also an astronomer, geodesist, and physicist; see, e.g., Mania (2018). Because of his outstanding scientific achievements, he was called Princeps Mathematicorum (Latin: the foremost of mathematicians) during his lifetime. As a nine-year-old, his teacher gave him the task of adding up the numbers 1 to 100. In the process, the young Gauss discovered this formula.

The Gaussian summation formula is usually covered in upper school level. It is fundamental and is needed in a wide variety of areas of mathematics. It is also useful for solving competitive mathematics problems for higher grades. For example, the Gaussian summation formula is needed in problem 540636 (state round, grade 6) from the 54th Mathematical Olympiad (Mathematik-Olympiaden e. V., 2015) for an intermediate step.

For clarity, the coin amounts are bracketed. On the board or in the students' note-books, the coin amounts can simply be circled. Then "MC" and the underlining of the coin values can be omitted, and the bracketing is then also no longer necessary.

The first two subtasks are quite simple and serve as an introduction. They should therefore be worked on and demonstrated by weaker students if possible.

(a) There are 5 ways to pay 8 MC with $(\underline{1}\text{-MC})$ and $(\underline{2}\text{-MC})$ coins:

$$0 \cdot (\underline{2}\text{-MC}) + 8 \cdot (\underline{1}\text{-MC}), 1 \cdot (\underline{2}\text{-MC}) + 6 \cdot (\underline{1}\text{-MC}), 2 \cdot (\underline{2}\text{-MC})$$
$$+ 4 \cdot (\underline{1}\text{-MC}), 3 \cdot (\underline{2}\text{-MC}) + 2 \cdot (\underline{1}\text{-MC}), 4 \cdot (\underline{2}\text{-MC}) + 0 \cdot (\underline{1}\text{-MC}).$$

(b) There are 7 ways to pay 13 MC:

$$0 \cdot (\underline{2}\text{-MC}) + 13 \cdot (\underline{1}\text{-MC}), 1 \cdot (\underline{2}\text{-MC}) + 11 \cdot (\underline{1}\text{-MC}), 2 \cdot (\underline{2}\text{-MC})$$
$$+ 9 \cdot (\underline{1}\text{-MC}), 3 \cdot (\underline{2}\text{-MC}) + 7 \cdot (\underline{1}\text{-MC}), 4 \cdot (\underline{2}\text{-MC}) + 5 \cdot (\underline{1}\text{-MC}),$$
$$5 \cdot (\underline{1}\text{-MC}) + 3 \cdot (\underline{1}\text{-MC}), 6 \cdot (\underline{2}\text{-MC}) + 1 \cdot (\underline{1}\text{-MC}).$$

There are 11 ways to pay 21 MC, namely.

$$0 \cdot (\underline{2}\text{-MC}) + 21 \cdot (\underline{1}\text{-MC}), 1 \cdot (\underline{2}\text{-MC}) + 19 \cdot (\underline{1}\text{-MC}), 2 \cdot (\underline{2}\text{-MC})$$
$$+ 17 \cdot (\underline{1}\text{-MC}), 3 \cdot (\underline{2}\text{-MC}) + 15 \cdot (\underline{1}\text{-MC}), 4 \cdot (\underline{2}\text{-MC})$$
$$+ 13 \cdot (\underline{1}\text{-MC}), 5 \cdot (\underline{2}\text{-MC}) + 11 \cdot (\underline{1}\text{-MC}), 6 \cdot (\underline{2}\text{-MC})$$
$$+ 9 \cdot (\underline{1}\text{-MC}), 7 \cdot (\underline{2}\text{-MC}) + 7 \cdot (\underline{1}\text{-MC}), 8 \cdot (\underline{2}\text{-MC})$$
$$+ 5 \cdot (\underline{1}\text{-MC}), 9 \cdot (\underline{2}\text{-MC}) + 3 \cdot (\underline{1}\text{-MC}), 10 \cdot (\underline{2}\text{-MC}) + 1 \cdot (\underline{1}\text{-MC}).$$

(c) **Didactic suggestion** To work on task part (c), (a) and (b) should be discussed again. If not already done, the payment possibilities should be sorted systematically as in the sample solution. For this purpose, it is recommended to sort unused coins as "$0 \cdot (\underline{1}\text{-MC})$" or "$0 \cdot (\underline{2}\text{-MC})$", even though this is of course superfluous from a mathematical point of view. Students should be guided to recognize that the number of $(\underline{2}\text{-MC})$ coins uniquely determines the number of $(\underline{1}\text{-MC})$ coins. So a payment option is already uniquely determined by the number of $(\underline{2}\text{-MC})$ coins.

Now let n be an even number. Then one can use at most (n: 2) many $(\underline{2}\text{-MC})$-coins, since with it the whole amount is already paid. So one can use 0, 1, 2, ... or (n: 2) many $(\underline{2}\text{-MC})$-coins. That is exactly (n: 2) + 1 possibilities.

If n is odd, we can use at most (n − 1): 2 many $(\underline{2}\text{-MC})$-coins, because then only 1 MC remains. So we can use 0, 1, 2, ... or (n − 1): 2 many $(\underline{2}\text{-MC})$ coins. That is exactly ((n − 1): 2) + 1 possibilities. So there are

$$A\left(n|\underline{1},\underline{2}\right) = \left(n:2\right) + 1 \text{ for even n} \tag{9.1}$$

$$A\left(n|\underline{1},\underline{2}\right) = \left(\left(n-1\right):2\right) + 1 \text{ for odd n.} \tag{9.2}$$

(d) It is $A\left(8|\underline{1},\underline{2}\right) = \left(8:2\right) + 1 = 4 + 1 = 5$, $A\left(13|\underline{1},\underline{2}\right) = \left(\left(13-1\right):2\right) + 1 = 6 + 1 = 7$

and $A\left(21|\underline{1},\underline{2}\right) = \left(\left(21-1\right):2\right) + 1 = 10 + 1 = 11$.

(e) It is $A\left(72|\underline{1},\underline{2}\right) = \left(72:2\right) + 1 = 36 + 1 = 37$ and

$A\left(53|\underline{1},\underline{2}\right) = \left(53-1\right):2 + 1 = 52:2 + 1 = 26 + 1 = 27.$

(f) is solved in (h)

(g) If Clemens uses two $(\underline{5}\text{-MC})$ coins, he still has 11 MC left to pay with $(\underline{1}\text{-MC})$ and $(\underline{2}\text{-MC})$ coins.   But   we   already   know   that   there   are

$A\left(11|\underline{1},\underline{2}\right)=\left((11-1):2\right)+1=6$ possibilities. If Clemens uses three ($\underline{5}$-MC) coins, there are 6 MC left, for which there are $A\left(6|\underline{1},\underline{2}\right)=\left(6:2\right)+1=4$ possibilities.

(h) To pay 21 MC, Clemens can use 0, 1, 2, 3, or 4 ($\underline{5}$-MC) coins. That leaves 21 MC, 16 MC, 11 MC, 6 MC, and 1 MC, respectively, to be paid for with ($\underline{1}$-MC) and ($\underline{2}$-MC) coins. The summands in the recursion formula Eq. 3.1 indicate in how many ways this is possible. With Eqs. 9.1 and 9.2 we get

$$
\begin{aligned}
A\left(21|\underline{1},\underline{2},\underline{5}\right) &= A\left(21|\underline{1},\underline{2}\right)+A\left(16|\underline{1},\underline{2}\right)+A\left(11|\underline{1},\underline{2}\right)+A\left(6|\underline{1},\underline{2}\right)+A\left(1|\underline{1},\underline{2}\right) \\
&= \left((21-1):2\right)+1+\left(16:2\right)+1+\left((11-1):2\right)+1+\left(6:2\right) \\
&\quad +1+\left((1-1):2\right)+1 \\
&= 10+1+8+1+5+1+3+1+0+1=31.
\end{aligned}
\tag{9.3}
$$

Thus, there are exactly 31 ways to pay 21 MC with ($\underline{1}$-MC), ($\underline{2}$-MC), and ($\underline{5}$-MC) coins. To increase readability, some parentheses have been placed in Eq. 9.3 and in other equations that are actually dispensable. In addition, this circumvents the dot before dash arithmetic rule, which primary school students do not yet know.

(i) To pay 19 MC, one can use 0, 1, 2, or 3 ($\underline{5}$-MC) coins. Analogous to h) one obtains

$$
\begin{aligned}
A\left(19|\underline{1},2,\underline{5}\right) &= A\left(19|\underline{1},2\right)+A\left(14|\underline{1},2\right)+A\left(9|\underline{1},2\right)+A\left(4|\underline{1},2\right) \\
&= \left((19-1):2\right)+1+\left(14:2\right)+1+\left((9-1):2\right)+1+\left(4:2\right)+1 \\
&= 9+1+7+1+4+1+2+1=26.
\end{aligned}
\tag{9.4}
$$

So there are 26 ways to pay 19 MC with ($\underline{1}$-MC), ($\underline{2}$-MC) and ($\underline{5}$-MC) coins.

(j) This subtask is even more complicated, because now it is possible to pay with four different coins instead of only three. Using the same idea that Clemens used to derive the formula Eq. 3.1 you get the recursion formula

$$
A\left(21|\underline{1},2,\underline{5},\underline{10}\right)=A\left(21|\underline{1},2,\underline{5}\right)+A\left(11|\underline{1},2,\underline{5}\right)+A\left(1|\underline{1},2,\underline{5}\right),
\tag{9.5}
$$

because one can use 0, 1 or 2 ($\underline{10}$-MC) coins. This leaves 21 MC, 11 MC, and 1 MC, respectively, to be paid with ($\underline{1}$-MC), ($\underline{2}$-MC), and ($\underline{5}$-MC) coins. This explains Eq. 9.5. From h) we already know that A $(21 \mid \underline{1}, \underline{2}, \underline{5}) = 31$. The terms A $(11 \mid \underline{1}, \underline{2}, \underline{5})$ and A $(1 \mid \underline{1}, \underline{2}, \underline{5})$ are simplified as we already know from h). Thus we get

$$
\begin{aligned}
A\left(21 \mid \underline{1}, \underline{2}, \underline{5}, \underline{10}\right) &= 31 + A\left(11 \mid \underline{1}, \underline{2}\right) + A\left(6 \mid \underline{1}, \underline{2}\right) + A\left(1 \mid \underline{1}, \underline{2}\right) + A\left(1 \mid \underline{1}, \underline{2}\right) \\
&= 31 + \left((11-1):2\right) + 1 + (6:2) + 1 + \left((1-1):2\right) + 1 + \left((1-1):2\right) + 1 \\
&= 31 + 5 + 1 + 3 + 1 + 0 + 1 + 0 + 1 = 43.
\end{aligned}
\tag{9.6}
$$

So there are 43 ways to pay 21 MC with ($\underline{1}$-MC), ($\underline{2}$-MC), ($\underline{5}$-MC) and ($\underline{10}$-MC) coins.

(k) Of course, there can be no sample solution for this subtask.

Didactic suggestion In k) each student can determine the level of difficulty by himself. The teacher can support. Students with a weaker performance should only consider ($\underline{1}$-MC) and ($\underline{2}$-MC) coins. Depending on the ability of the students, the instructor can omit tasks j) and k).

Mathematical Goals and Outlook
Students learn how to derive a recursion formula to reduce a complicated mathematical problem to simpler subproblems and eventually solve it. Students have already learned about reducing to simpler problems in games and word puzzles in Volume I (Schindler-Tschirner & Schindler, 2019).

Recursion formulas occur in mathematics (for example, to define the Fibonacci sequence, cf. e.g., (Schiemann & Wöstenfeld, 2017), p. 101 f. and computer science (e.g., recursive functions) in different contexts; cf. e.g. (Kreß, 2004), for example.

Several terms are introduced in this chapter (divisor, prime number, prime factorization, square number). Although at least "divisor" and "prime factorization" are intuitive, and prime numbers are probably already known by some of the students, sufficient time should be allowed for students to become familiar with the terms.

(a) The subtask (a) is easy to understand. It serves to familiarize the children with the definition of a divisor.

Didactic suggestion In order not to need too much time at this point, nor to allow boredom to arise, the children can be divided into two or three groups, with each group independently determining the divisors of some numbers between 1 and 30; in the case of two groups, for example, for the range 1 to 15 or for 16 to 30; in the case of three groups, for example, for 1 to 10, 11 to 20 or 21 to 30. This creates the first sense of achievement. The children can then present their solutions on the board. In individual lessons, the teacher takes over some numbers himself. Here it is recommended to form "more interesting" sets of numbers, e.g. {1, 2, 6, 9, ..., 29} and the remaining numbers, so that the mutual presentation of the solutions does not become too boring for the child. (Of course, this is also an option for groups of students).

In the following, the divisors of the numbers 1 to 30 are given without further comment. To save typing, T(n) denotes the set of all divisors of n. The set notation can be circumvented by entering the divisors, e.g., in a two-column table, where the left column contains the number and the right column its divisors.

$$T(1) = \{1\}, T(2) = \{1, 2\}, T(3) = \{1, 3\}, T(4) = \{1, 2, 4\}, T(5) = \{1, 5\},$$
$$T(6) = \{1, 2, 3, 6\}, T(7) = \{1, 7\}, T(8) = \{1, 2, 4, 8\}, T(9) = \{1, 3, 9\},$$
$$T(10) = \{1, 2, 5, 10\}, T(11) = \{1, 11\}, T(12) = \{1, 2, 3, 4, 6, 12\},$$
$$T(13) = \{1, 13\}, T(14) = \{1, 2, 7, 14\}, T(15) = \{1, 3, 5, 15\},$$
$$T(16) = \{1, 2, 4, 8, 16\}, T(17) = \{1, 17\}, T(18) = \{1, 2, 3, 6, 9, 18\},$$
$$T(19) = \{1, 19\}, T(20) = \{1, 2, 4, 5, 10, 20\}, T(21) = \{1, 3, 7, 21\},$$
$$T(22) = \{1, 2, 11, 22\}, T(23) = \{1, 23\}, T(24) = \{1, 2, 3, 4, 6, 8, 12, 24\},$$
$$T(25) = \{1, 5, 25\}, T(26) = \{1, 2, 13, 26\}, T(27) = \{1, 3, 9, 27\},$$
$$T(28) = \{1, 2, 4, 7, 14, 28\}, T(29) = \{1, 29\}, T(30) = \{1, 2, 3, 5, 6, 10, 15, 30\}.$$

(b) The number 1 has only one divisor. The numbers 24 and 30 have the most divisors, 8. Exactly two divisors have the numbers 2, 3, 5, 7, 11, 13, 17, 19, 23, and 29.

(c) Fourth graders should know prime numbers from school lessons. Subtasks (c) and (d) are designed to help students recall prime numbers or become familiar with them for the first time.

Therefore, the subtasks (c) and (d) should be given some time. Of course, there can be no model solution for (c). The prime numbers between 1 and 100 are: 2, 3, 5, 7, 11, 13, 17, 19, 23, 29, 31, 37, 41, 43, 47, 53, 59, 61, 67, 71, 73, 79, 83, 89, 93, 97.

(d) Prime numbers are 7, 41 and 83. The other numbers are not prime numbers:
$14 = 2 \cdot 7$, $51 = 3 \cdot 17$, $72 = 8 \cdot 9$ (or $2 \cdot 36$ or $3 \cdot 24$ etc.), $100 = 10 \cdot 10$.

(e) If we use the preliminary work from (a) and (b), we don't have to calculate any more: The numbers 2, 3, 5, 7, 11, 13, 17, 19, 23 and 29 are the prime numbers between 1 and 30. From the definition of a prime number (divisible only by 1 and itself) it follows, after all, that the prime numbers are those numbers with (exactly) two divisors.

$$2 = 2, \ 3 = 3, \ 4 = 2^2, \ 5 = 5, \ 6 = 2 \cdot 3, \ 7 = 7, \ 8 = 2^3, \ 9 = 3^2, \ 10 = 2 \cdot 5,$$
$$11 = 11, \ 12 = 2^2 \cdot 3, \ 13 = 13, \ 14 = 2 \cdot 7, \ 15 = 3 \cdot 5.$$

Note Of course, $4 = 2 \cdot 2$ etc. is also correct. The power notation will be explained in the next chapter, in order not to overload this chapter with too many explanations (definitions). Of course, the power notation can also be used already in this chapter.

Note The instructor should definitely point out that the prime factors can be determined step by step.

Example $12 = 2 \cdot 6 = 2 \cdot 2 \cdot 3$.

$$16 = 2^4, \ 17 = 17, \ 18 = 2 \cdot 3^2, \ 19 = 19, \ 20 = 2^2 \cdot 5, \ 21 = 3 \cdot 7,$$
$$22 = 2 \cdot 11, \ 23 = 23, \ 24 = 2^3 \cdot 3, \ 25 = 5^2, \ 26 = 2 \cdot 13, \ 27 = 3^3,$$
$$28 = 2^2 \cdot 7, \ 29 = 29, \ 30 = 2 \cdot 3 \cdot 5.$$

(f) Of the numbers 1 to 30, only 1, 4, 9, 16, and 25 have an odd number of divisors. These are exactly the square numbers that are not greater than 30.

(g) From (h), we can conjecture that of the numbers up to 200, only the square numbers, i.e., 1, 4, 9, 16, 25, 36, 49, 64, 81, 100, 121, 144, 169, and 196, have an odd number of divisors. This conjecture can be randomly tested by the students for individual numbers.

(h) In this subtask, we prove the conjecture about the square numbers from subtask (i). The proof uses only simple mathematical tools, but experience shows that it is not easy to understand, at least for third graders. The instructor may choose to omit this subtask, taking into account the proficiency level of the course.

Preliminary Remark The proof idea is reminiscent of kindergarten children lining up in rows of two (which was common, at least in the past), with the children who are in the same row holding hands. If the number of children is odd, there is only a single child in one row. Otherwise, all rows are occupied by two children each.

Proof Let n be a natural number and a be a divisor of n. Then there is a natural number b for which $n = a \cdot b$, namely $b = n : a$ (Example: $n = 10$, $a = 2$. Here $b = 10 : 2 = 5$.) The number b is a divisor of n as well. For the moment we call the number b the "partner" of the number a. But then a is also the (only) "partner" of b.

If a and b were children, they would be in the same row. In this way, we can assign a uniquely determined partner to each divisor of n, and no divisor occurs in more than one such pair. However, there is one exceptional case to consider, namely when a = b, i.e. n = a · a. But this can only happen if n is a square number, and even then only for exactly one divisor (the root of n, but this need not be discussed here). So the pairs normally consist of two different numbers; only in the case of square numbers there is a number which gives itself the hand, so to speak. This shows that square numbers always have an odd number of divisors, while non-quadratic numbers have an even number of divisors.

Didactic Suggestion   Figure 10.1 illustrates the proof idea for the numbers 12 and 16. Before the general proof, the teacher can illustrate the proof idea on the numbers 12 and 16.

Mathematical Goals and Outlook
Prime numbers and the divisibility of natural numbers play an important role in mathematics. To some extent, such questions are also dealt with in school lessons, for example to determine the greatest common divisor or the least common multiple of natural numbers (usually in the lower school level). In mathematical terms, the explanations under "Dividus explains" are called definitions. In the last subtask, a mathematical proof is given again.

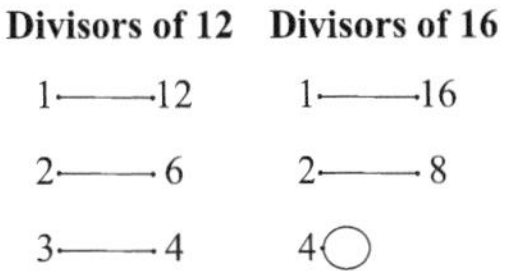

Fig. 10.1   Divisors of 12 and 16 in rows of two

Divisors of 12	Divisors of 16
1———12	1———16
2——— 6	2——— 8
3——— 4	4◯

Sample Solution for Chapter 5 11

This mathematical adventure is exceptionally comprehensive and, in terms of content, certainly the most challenging in this volume. Two or even three teaching units should be used for this mathematical adventure.

Didactic Suggestion Especially for younger students, this chapter presents a great challenge. The instructor is free to omit subtasks (f) and (g) and to explain the calculation formulas Eqs. 11.5, 11.6 and 11.7 to the students using examples, without deriving them.

The first two exercises are relatively simple and the procedure is already known from the last chapter. Therefore, all children should be able to solve them and gain their first sense of achievement.

(a) $63 = 3 \cdot 21 = 3 \cdot 3 \cdot 7$

(b) $125 = 5 \cdot 25 = 5 \cdot 5 \cdot 5$

(c) Dividus explains the power notation, even though some children may already know this from the lessons

$$63 = 3^2 \cdot 7 \quad \text{und} \quad 125 = 5^3$$

© The Author(s), under exclusive license to Springer Fachmedien Wiesbaden GmbH, part of Springer Nature 2023
S. Schindler-Tschirner, W. Schindler, *Mathematical Stories II - Recursion, Divisibility and Proofs*, essentials,
https://doi.org/10.1007/978-3-658-38611-5_11

It is possible that the students have already used the power notation in subtasks (a) and (b). In that case, subtask (c) has already been solved.

What should Clemens notice when he looks at the prime factorization of 12?

Observation   In the prime factorizations of the divisors of 12 no other prime factors occur than in the prime factorization of $12 = 2^2 \cdot 3^1$ (i.e. 2 and 3). No prime factor occurs more frequently than in the prime factorization of 12. More precisely, any product $2^s \cdot 3^t$ is a divisor of 12 if $s \in \{0, 1, 2\}$ and $t \in \{0, 1\}$.

Explanation   Suppose a is a divisor of 12. Then $n = a \cdot b$ for $b = 12$: a. If $a = 1$ or $a = 12$, a has the above form. Otherwise, the prime factorization of 12 can be determined stepwise, that is, for a and b separately. Therefore, $a = 2^s \cdot 3^t$ must be, and the exponents s and t cannot be greater than 2 and 1, respectively. On the other hand, any number $a = 2^s \cdot 3^t$ is a divisor of 12 if s and t are not greater than 2 and 1, respectively; namely, then $12 = a \cdot b$ is valid for $b = 2^{2-s} \cdot 3^{1-t}$, since one may interchange the factors when multiplying. The number b is the product of the "remaining" prime factors.

Addition (for the instructor)   Using the same reasoning, we can show that the above observation holds in general for any number n. The divisors of n can be represented as the product of powers of prime factors occurring in the prime factorization of n. Here the exponents must not be larger than in the prime factorization of n (0 is possible). Conversely, all combinations of exponents allowed in this respect yield divisors of n. This is the key to the solution.

Didactic Suggestion   Equations 5.1 and 5.2, subtasks (d) and (e) and later Eq. 5.3, should lead the children to the realization how the divisors of a number can be described as a product of prime powers. Possibly the children (unlike Clemens) already recognize the regularity after (d) or (e). In that case, the teacher can bring forward subtask (h) (initially without the calculation part). The description of the divisors is important, while the reasoning can be treated shorter if necessary.

As described in Chap. 10 T(n) denotes the set of all divisors of the number n. It is up to the instructor whether to use this shorthand notation or to write "divisors of n" at some length, as in the assignment.

(d) $20 = 2 \cdot 10 = 2 \cdot 2 \cdot 5 = 2^2 \cdot 5,\ T(20) = \{1,\ 2,\ 4,\ 5,\ 10,\ 20\}$
    $= \left\{1,\ 2,\ 2^2,\ 5,\ 2 \cdot 5,\ 2^2 \cdot 5\right\},\ 6$ divisors

(e) $35 = 5 \cdot 7,\ T(35) = \{1, 5, 7, 5 \cdot 7\},\ 4$ divisors

In (d) and (e) we have enumerated the divisors of the numbers 20 and 35 in power notation. The number of divisors of a number n does not depend on its prime factors themselves, but only on how many prime factors and in which power these prime factors occur in the prime factorization of n. One can easily convince oneself that, for example, both $6 = 2 \cdot 3$ and $35 = 5 \cdot 7$ have 4 divisors each. The task of determining the number of divisors of a number is reduced to a simple combinatorial problem. The subtasks (f) and (g), which introduce elementary combinatorics, serve as preparation.

(f) **Observation** The outfit of Ron Rodent is described by the choice of the shirt {b,y,r} and the trousers {st,do}. For example, one possible combination is (y,st). This means that Ron Rodent puts on his yellow shirt and his striped pants.

What is wanted is the number of all possible clothing combinations of shirt and pants. The teacher should first give the children the opportunity to write down all possible combinations. Perhaps some of the children will already recognize the law they are looking for.

Let's now turn to the systematic solution: Ron can choose from three shirts and two pairs of pants. He can wear either the striped or the dotted pants with the blue shirt. These are two possibilities. Of course, the same is true for the yellow and the red shirt, because the choice of pants is independent of the shirt color. So in total Ron Rodent has $3 \cdot 2 = 6$ possibilities to combine his shirts with the pants.

(g) **Observation** In (g) Ron's socks {b,w,ch,p} are also considered. A possible combination now consists of three garments rather than two. For example, (b,do,st) means that Ron Rodent puts on his blue shirt, his dotted pants, and his checkered socks.

Again, the children should first write down and collect all combinations, unless the general regularity has already been identified in (f).

Obviously, Ron can choose four pairs of socks for each combination of shirt and pants (e.g., red shirt with striped pants). But we already know from (f) that there are $3 \cdot 2 = 6$ combinations of shirt and pants. It follows that Ron Rodent has a total of $3 \cdot 2 \cdot 4 = 24$ combinations of shirt, pants, and socks.

Note that 24 is the product of the number of shirts (=3), the number of pants (=2), and the number of socks (=4). So Ron Rodent can dress differently on 24 consecutive days!

Note   In combinatorics, such problems are often described by urn models. For (g) this means: There are three urns. In the first urn there are three balls, labeled "b", "y" and "r". In the second urn there are two balls ("st" and "do"), while in the third urn there are four balls ("b", "w", "ch" and "p"). If you draw one ball from each urn, this determines the clothes of Ron Rodent. The total number of possible clothing combinations is obtained by multiplying the number of balls that are in each urn (here: $3 \cdot 2 \cdot 4 = 24$).

We need these considerations to solve the following subtasks.

(h) As a reminder, $12 = 2^2 \cdot 3^1$. In Eq. 5.3 the sought-after regularity appears openly (cf. Observation, Explanation and Addition). The divisors of 12 are always represented as a product of powers of 2 *and* 3, even if one or both exponents are 0. (It is then no longer a prime factorization, since "1" appears as a factor, but that is not important here). The divisors of 12 can be described by the six pairs (0,0), (0,1), (1,0), (1,1), (2,0), (2,1). The first number of a bracket corresponds to the exponent of the first prime factor in the prime factorization (here: 2) and the second number of this bracket corresponds to the exponent of the second prime factor (here: 3). Table 11.1 illustrates the correspondence

Table 11.1   The divisors of 12 and 35

$12 = 2^2 \cdot 3^1$		$35 = 5^1 \cdot 7^1$	
Pairs	Divider	Pairs	Divider
(0,0)	$2^0 \cdot 3^0 = 1$	(0,0)	$5^0 \cdot 7^0 = 1$
(0,1)	$2^0 \cdot 3^1 = 3$	(0,1)	$5^0 \cdot 7^1 = 7$
(1,0)	$2^1 \cdot 3^0 = 2$	(1,0)	$5^1 \cdot 7^0 = 5$
(1,1)	$2^1 \cdot 3^1 = 6$	(1,1)	$5^1 \cdot 7^1 = 35$
(2,0)	$2^2 \cdot 3^0 = 4$		
(2,1)	$2^2 \cdot 3^1 = 12$		

Correspondence between pairs of numbers and divisors

between pairs of numbers and the divisors using the numbers 12 and 35 as examples.

Observation This behaves exactly as with Ron Rodent's shirts and pants in sub-task (f): There Ron had 3 shirts (here: first exponent = 0, 1 or 2) and 2 pants (here: second exponent = 0 or 1) to combine. So $12 = 2^2 \cdot 3^1$ has exactly $3 \cdot 2 = 6$ divisors. Or expressed differently:

$$\big((\text{exponent of } 2) + 1\big) \cdot \big((\text{exponent of } 3) + 1\big) = 6 \qquad (11.1)$$

Or in the urn model: There are two urns, where in the first urn there are three red balls with the inscription "0", "1" and "2" and in a second urn there are two green balls with the inscription "0" and "1". How many possibilities are there if you draw exactly one ball from each urn?

The teacher can now go directly to (i) or, for practice, first work out with the children the number of divisors of 20 and of 35.

Solution: $20 = 2^2 \cdot 5^1$ has a total of $(2 + 1) \cdot (1 + 1) = 6$ divisors, but $35 = 5^1 \cdot 7^1$ has $(1 + 1) \cdot (1 + 1)$ divisors.

(i) It is $55 = 5^1 \cdot 11^1$. Therefore, 55 has a total of $(1 + 1) \cdot (1 + 1) = 4$ divisors. (It is $T(55) = \{1, 5, 11, 55\} = \{5^0 \cdot 11^0, 5^1 \cdot 11^0, 5^0 \cdot 11^1, 5^1 \cdot 11^1\}$.)

Observation The prime factors themselves are not important, but only how many different prime factors occur how often in the prime factorization of a number n. If we look at their divisors, each single prime factor can occur there at most as often as in the prime factorization of n itself (= exponent of this prime number in the prime factorization of n), and everything between 0 and this number is possible!

Some more exercises follow, where the prime factorization and the facts just learned can be practiced. It is up to the instructor to divide the tasks into groups or to omit some.

(j) $100 = 2^2 \cdot 5^2$. Therefore, 100 has a total of. $(2 + 1) \cdot (2 + 1) = 9$ divisors. As explained in detail above, the summand " + 1" results from the fact that the exponents 0, 1, and 2 are possible. The remaining subtasks are solved analogously.

(k) $99 = 3^2 \cdot 11^1$. Therefore 99 has a total of $(2 + 1) \cdot (1 + 1) = 3 \cdot 2 = 6$.

(l) $128 = 2^7$. Therefore, 128 has a total of $(7 + 1) = 8$ divisors.

(m) $168 = 2^3 \cdot 3^1 \cdot 7^1$. Here for the first time three different prime factors occur (counterpart to (g)). Therefore 168 has a total of

$$((\text{exponent of } 2) + 1) \cdot ((\text{exponent of } 3) + 1) \cdot ((\text{exponent of } 7) + 1)$$
$$= (3 + 1) \cdot (1 + 1) \cdot (1 + 1) = 4 \cdot 2 \cdot 2 = 16 \text{ divisors} \tag{11.2}$$

(n) $525 = 3^1 \cdot 5^2 \cdot 7^1$. Therefore 525 has a total of $(1 + 1) \cdot (2 + 1) \cdot (1 + 1) = 2 \cdot 3 \cdot 2 = 12$ divisors.

(o) $529 = 23^2$. Since 23 is a prime number, 529 has only 3 divisors.

In each of the subtasks (m) and (n) three different prime factors occur. We can imagine that there are blue balls in a third urn, which indicate how often the prime factor 7 occurs. Normally, large numbers have many divisors, but 529 shows that this is not always the case.

Additions   Our calculation formula can be generalized to any natural number n:

$$Let\ n = p_1^{c_1} \cdot p_2^{c_2} \cdots \cdot p_k^{c_k} \ (\text{prime factorization of } n) \cdot \tag{11.3}$$

Here p_1, p_2, ..., pk denote distinct primes, and the exponents c_1, c_2, ..., ck are greater than or equal to 1. Then holds (see, e.g., (Menzer & Althöfer, 2014), Theorem 4.2.1):

The integer $n = p_1^{c_1} \cdot p_2^{c_2} \cdots \cdot p_k^{c_k}$ has $(c_1 + 1) \cdots (c_k + 1)$ divisors. $\tag{11.4}$

Example $12 = 2^2 \cdot 3$. Here $p_1 = 2$, $c_1 = 2$, $p_2 = 3$, and $c_2 = 1$.

$525 = 3^1 \cdot 5^2 \cdot 7^1$. Here is $p_1 = 3$, $c_1 = 1$, $p_2 = 5$, $c_2 = 2$, $p_3 = 7$, and $c_3 = 1$.

Equation 11.4 is already rather "burdened with indices". Therefore, the (well-known) general formula Eq. 11.4 is intended only for the instructor to answer questions from students about how it behaves when the prime factorization of n contains more than three different prime factors. This is likely to be too difficult for students.

In particular, Eq. 11.4 gives the calculation formulas for the special cases where there are one, two, or three different prime factors in the prime factorization of n. In Eqs. 11.5, 11.6 and 11.7 indices were omitted.

$$\text{The number } n = p^s \text{ has } (s+1) \text{ divisors.} \tag{11.5}$$

$$\text{The number } n = p^s \cdot q^t \text{ has } (s+1)\cdot(t+1) \text{ divisors.} \tag{11.6}$$

$$\text{The number } n = p^s \cdot q^t \cdot r^u \text{ has } (s+1)\cdot(t+1)\cdot(u+1) \text{ divisors.} \tag{11.7}$$

Here (p and q) or (p, q, and r) denote distinct prime factors, and the exponents s, t, and u are greater than or equal to 1.

Mathematical Goals and Outlook

First, the important technique of prime factorization is practiced again by several tasks, as it plays an important role in mathematics (cf. Chap. 10). In this mathematical adventure, the children are guided by selected examples to the solution of the initial problem (number of divisors), which at first glance seems to have little to do with prime numbers.

This also requires elementary combinatorial considerations, which are also used in tasks of various mathematics competitions, e.g. the Mathematics Olympiad, for elementary school (cf. e.g., (Mathematik-Olympiaden e. V. 2013), tasks 470412, 500414, 520321, 520411 (grades 3 and 4)) or lower school level. In the classroom, combinatorics (with more in-depth content) as a subfield of stochastics is usually not on the curriculum until upper school level.

Using the general calculation formula Eq. 11.4, one can, for example, solve (with a simple additional consideration) the problem 561234 (state round, grade 12/13) from the 56th Mathematical Olympiad (Mathematik-Olympiade e. V., 2017).

Sample Solution for Chapter 6   12

After the very demanding Chap. 5 the last two chapters are much easier, because there we basically calculate.

(a) 16: 7 = 2 remainder 2, 9: 7 = 1 remainder 2, 2: 7 = 0 remainder 2, 70: 7 = 10 remainder 0

(b) 16 : 5 = 3 remainder 1,   11 : 5 = 2 remainder 1, 9 : 5 = 1 remainder 4

Note on the Definition of a ≡ b mod n   The numbers a and b are integers, so they can take values in the set $Z = \{ \cdots -3, -2, -1, 0, 1, 2, 3, \cdots \}$ can take on values in the set. However, in the tasks, a and b are never negative because elementary students do not yet know negative numbers.

Didactic Suggestion   The teacher can briefly explain to the students that the non-negative integers 0, 1, 2, … are a subset of the integers and that there are other integers (negative numbers), but that only non-negative numbers are used in the SG. The students are familiar with these. Alternatively, the instructor could include the non-negativity of a and b in the definition. However, this would then no longer correspond to the usual definition, and for the "Supplementary remarks" in Chap. 13 this definition would then have to be extended to integers again.

The modulo arithmetic is based on the division with remainder, which is taught in primary school. However, with the modulo arithmetic, one is only interested in the division remainder.

(c)
$$22 \equiv 2 \bmod 10, \quad 171 \equiv 1 \bmod 2, \quad 22 \equiv 7 \bmod 15,$$
$$52 \equiv 2 \bmod 25, \quad 17 \equiv 3 \bmod 7, \quad 22 \equiv 22 \bmod 28.$$

Note Of course, $22 \equiv 12 \bmod 10$ is also correct, but 12 is not the smallest nonnegative number that satisfies this congruence. Usually the smallest solutions are of particular interest, as we will see below (e.g., in the context of time-of-day tasks).

In the tasks concerning the days of the week, the number 7 plays the central role, because there are 7 days of the week that repeat. In the case of the time, the number 24 takes over the role of 7, because the day has 24 h and therefore after 24 h it is the same time as right now. If one uses only 1 to 12 o'clock (in the morning as in the afternoon), then the modulus 12 is relevant.

(d) It is $26 \equiv 2 \bmod 24$. Therefore, it is the same time in 26 h as in 2 h. Since it is now 18 o'clock (6 pm), it is then 20 o'clock (8 pm).

(e) It is $52 \equiv 4 \bmod 24$. Therefore, it is the same time in 52 h as it is in 4 h. Since it is now 10 o'clock (10 am), it is then 14 o'clock (2 pm).

(f) It is $27 \equiv 3 \bmod 24$. Therefore, it is as late in 27 h as it is in 3 h. Since it is now 23 o'clock (11 pm), it is then 2 o'clock (2 am).

(g) $29 \equiv 5 \bmod 24, \quad 241 \equiv 1 \bmod 24, \quad 59 \equiv 11 \bmod 24,$

(h) Let us use the modulo calculation for this as well. There is exactly 1 year between January 1, 2019 and January 1, 2020. The year 2019 is not a leap year and therefore has 365 days.
Now 365: 7 = 52 remainder 1, so $365 \equiv 1 \bmod 7$. So January 1, 2020 is a Wednesday.

(i) There are exactly 4 years between January 1, 2019 and January 1, 2023, with 2020 being a leap year and therefore having 366 days. Therefore, a total of 36
5 + 366 + 365 + 365 = 1461 days will pass. Now 1461: 7 = 208 remainder 5, so $1461 \equiv 5 \bmod 7$, and January 1, 2023 is a Sunday. Velox had not considered in his answer that 2020 is a leap year.

Mathematical Goals and Outlook
Calculating with remainders is first motivated by day-of-week and time-of-day problems. Then the modular arithmetic (modulo calculation) is formally introduced and practiced on several examples. Modular calculation is further deepened in Chap. 7. We will get to know calculation rules which simplify the practical application of the modulo calculation considerably.

The first exercises are again relatively easy, but it is very important that the children internalize the calculation rules. With calculation rule 1 you get

(a) $22 + 17 \equiv 2 + 7 \equiv 9 \bmod 10,$ $100 + 17 \equiv 0 + 7 \equiv 7 \bmod 10,$

           $31 + 17 \equiv 1 + 2 \equiv 3 \equiv 0 \bmod 3,$ $7 + 2 \equiv 3 + 2 \equiv 5 \equiv 1 \bmod 4,$

           $12 + 2 + 3 \equiv 0 + 0 + 1 \equiv 1 \bmod 2.$

Note The numbering of the modulo calculation rules is not common in the literature, but serves here only the shorter designation.

(b) From subtask h) from Chap. 6 we already know that $365 \equiv 1 \bmod 7$ is valid. From this it now follows without any major calculation:

$$365 + 366 + 365 + 365 \equiv 1 + 2 + 1 + 1 \equiv 5 \bmod 7.$$

January 1, 2023 is therefore a Sunday.

(c) $22 \cdot 22 \equiv 1 \cdot 1 \equiv 1 \bmod 7,$ $10 \cdot 17 \equiv 1 \cdot 2 \equiv 2 \bmod 3,$

           $31 \cdot 17 \equiv 0 \cdot 17 \equiv 0 \bmod 31.$

© The Author(s), under exclusive license to Springer Fachmedien
Wiesbaden GmbH, part of Springer Nature 2023
S. Schindler-Tschirner, W. Schindler, *Mathematical Stories II - Recursion,
Divisibility and Proofs*, essentials,
https://doi.org/10.1007/978-3-658-38611-5_13

(d) It is $10 \equiv 1$ mod 3. If we use this result and apply the calculation rule 2, we get $100 = 10 \cdot 10 \equiv 1 \cdot 1 \equiv 1$ mod 3. In the same way, one shows $1000 \equiv 10 \cdot 100 \equiv 1 \cdot 1 \equiv 1$ mod 3.

Likewise

$$10 \equiv 1 \text{ mod } 9, \qquad 100 = 10 \cdot 10 \equiv 1 \cdot 1 \equiv 1 \text{ mod } 9 \text{ and}$$
$$1000 = 10 \cdot 100 \equiv 1 \cdot 1 \equiv 1 \text{ mod } 9.$$

(e) Using calculation rule 2 and the results from d) we get

$$3000 = 3 \cdot 1000 \equiv 3 \cdot 1 \equiv 3 \text{ mod } 9, \qquad 200 = 2 \cdot 100 \equiv 2 \cdot 1 \equiv 2 \text{ mod } 9,$$
$$40 = 4 \cdot 10 \equiv 4 \cdot 1 \equiv 4 \text{ mod } 9.$$

(f) With arithmetic rule 1 and subtask e) so follows immediately

$$3246 = 3000 + 200 + 40 + 6 \equiv 3 + 2 + 4 + 6 \equiv 15 \equiv 6 \text{ mod } 9.$$

So the number 3246 has the remainder 6 modulo 9.

(g) In the same way as in f) one calculates

$$3564 = 3000 + 500 + 60 + 4 = 3 \cdot 1000 + 5 \cdot 100 + 6 \cdot 10 + 4$$
$$\equiv 3 \cdot 1 + 5 \cdot 1 + 6 \cdot 1 + 4 \equiv 3 + 5 + 6 + 4 \equiv 18 \equiv 0 \text{ mod } 9.$$

So the number 3564 is divisible by 9.

(h) first multiplication task: using the digit sum rule, one immediately obtains $34 \cdot 54 \equiv 7 \cdot 0 \equiv 0 \text{ mod } 9$ but $1736 \equiv 1 + 7 + 3 + 6 \equiv 17 \equiv 8 \text{ mod } 9$.

Of course, if the result of the multiplication were correct, both sides of the multiplication task would have to have the same remainder modulo 9. (In fact $34 \cdot 54 = 1836$.)

second multiplication task: The same is true for $27 \cdot 44 \equiv 0 \cdot 8 \equiv 0 \mod 9$ but $1178 \equiv 1 + 1 + 7 + 8 \equiv 17 \equiv 8 \mod 9$. So this multiplication task is also wrong. (In fact $27 \cdot 44 = 1188$.)

third multiplication task: It is $24 \cdot 19 \equiv 6 \cdot 1 \equiv 6 \mod 9$ and likewise is $456 \equiv 4 + 5 + 6 \equiv 15 \equiv 6 \mod 9$.

fourth multiplication task: Here is $37 \cdot 41 \equiv 1 \cdot 5 \equiv 5 \mod 9$ and also $1508 \equiv 1 + 5 + 0 + 8 \equiv 14 \equiv 5 \mod 9$.

So the third and the fourth result *could be* correct. In fact, only the third result is correct, while $37 \cdot 41 = 1517$.

The procedure in subtask h) is also known as the "modulo 9 remainder test" and was a tried and tested method for discovering calculation errors in written multiplications before the development of calculators. However, the instructor must make it clear here that the modulo 9 remainder test can only be used to determine *(with certainty)* that a calculation result is incorrect, but not that it is correct. If the product of the remainders modulo 9 of the factors matches with the remainder modulo 9 of the calculated result (as in the third and fourth multiplication task), the result *can be* correct (third multiplication task), but it does not *have to be* (fourth multiplication task).

Supplementary remarks   In the explanations of the last mathematical adventure, it was already pointed out that the modulo calculation does not only apply to the natural numbers, but includes the negative integers. Since this *essential* is aimed at primary school children, the negative numbers are left out. By analogy with the calculation rules for addition and multiplication, the following also applies.

Modulo calculation rule 3 (subtraction) $a \equiv a' \mod n$ and $b \equiv b' \mod n$ imply $a - b \equiv a' - b' \mod n$.

Example   It is $19 \equiv 9 \mod 10$ and $12 \equiv 2 \mod 10$. From calculation rule 3, it follows $19 - 12 \equiv 9 - 2 \equiv 7 \mod 10$.

It may happen that intermediate results are negative. Then you can simply add the modulus until the result is greater than or equal than 0.

Example   It is $22 \equiv 2 \mod 10$ and $19 \equiv 9 \mod 10$. From calculation rule 3 it follows $22 - 19 \equiv 2 - 9 \equiv -7 \equiv -7 + 10 \equiv 3 \mod 10$.

Calculation rule 3 can be used to solve other interesting problems. Several tasks are addressed and solved below. If lower level students participate, the instructor can explain calculation rule 3 and add these tasks as well.

Additional Task_a)   Determine the smallest nonnegative number for which the congruence is correct. Calculate skillfully:

$$242 - 111 \equiv \quad \text{mod } 10, \quad\quad 100 - 17 \equiv \quad \text{mod } 10, \quad\quad 301 - 17 \equiv \quad \text{mod } 3.$$

Solution:

$$242 - 111 \equiv 2 - 1 \equiv 1 \text{ mod } 10, \quad\quad 100 - 17 \equiv 0 - 7 \equiv -7 + 10 \equiv 3 \text{ mod } 10,$$
$$301 - 17 \equiv 1 - 2 \equiv -1 \equiv 3 - 1 \equiv 2 \text{ mod } 3.$$

Additional Task_b)   Work out which day of the week you were born on.
Solution: we explain the problem with an example. Ben celebrated his tenth birthday on May 18, 2018. A look at the calendar of the year 2018 shows that this is a Friday. Exactly 10 years have passed since his birth. Ignoring leap days for the moment, this is $10 \cdot 365$ days. In addition, there are 2 leap days (February 29, 2012, February 29, 2016). Since the birth is in the past, this results in a negative sign. As we already know, $365 \equiv 1 \text{ mod } 7$, and thus follows altogether $-(10 \cdot 365 + 2) \equiv -(3 \cdot 1 + 2) \equiv -5 \equiv 7 - 5 \equiv 2 \text{ mod } 7$. So Ben was born on a Sunday.

Mathematical Goals and Outlook
In the last mathematical adventure the modular arithmetic is deepened. Useful calculation rules for addition and multiplication are introduced, which significantly expand the application areas of modulo calculations.

Modular arithmetic plays an important role in number theory. With its help, for example, one can solve questions about divisibilities, derive and prove divisibility rules (cf., e.g., the note of Dwarf Modulus on the divisibility rules for the numbers 3 and 9), and sometimes one can prove the nonexistence of solutions. Nevertheless, modular arithmetic is hardly treated in school. An introduction for (older) students to modulo calculation including exercises can be found e.g. in (Meier, 2003), Chap. 3.
Modulus calculation is also extremely useful for many tasks in mathematics competitions, such as the Mathematics Olympiad or the Federal Mathematics Competition. An example of this is task 491331 (state round, grades 12/13) from

the 49th Mathematical Olympiad (Mathematik-Olympiade e. V. 2010). Using modulo, it is much easier to show that the eq. $2010\,x^2 - 2009\,y^2 = 50$ has no integer solutions x and y than in the sample solution. With the calculation rules 1, 2 and 3 and the digit sum rule (applied to 2009 and 2010) we get

$$2010 \cdot x^2 - 2009 \cdot y^2 \equiv 0 \cdot x^2 - 2 \cdot y^2 \equiv 3 \cdot y^2 - 2 \cdot y^2 \equiv y^2 \equiv 50 \equiv 2 \bmod 3. \qquad (13.1)$$

With arithmetic rule 2 it is easy to show that square numbers can have only a remainder modulo 3 of 0 or 1. So the congruence $y^2 \equiv 2 \bmod 3$ is not solvable.

Modulo calculation also plays an important role in cryptography, for example in the widely used RSA algorithm (Beutelspacher, 2015). In the RSA algorithm, very large numbers are exponentiated, but only the remainders with respect to a (very large) modulus are calculated.

What You Learned from This *Essential*

This book provides carefully designed learning units with detailed sample solutions for a mathematics SG for gifted students in primary school. In six mathematical stories, you have

- learned about the Gaussian summation formula and practiced it with examples.
- solved a difficult problem recursively.
- worked on elementary combinatorics problems.
- derived how to calculate the number of divisors of a number from its prime factorization.
- learned about and applied modular arithmetic (modulo calculation).
- learned that proofs are necessary in mathematics, and you have provided proofs yourself in different contexts.

© The Author(s), under exclusive license to Springer Fachmedien
Wiesbaden GmbH, part of Springer Nature 2023
S. Schindler-Tschirner, W. Schindler, *Mathematical Stories II - Recursion,
Divisibility and Proofs*, essentials,
https://doi.org/10.1007/978-3-658-38611-5

References

Amann, F. (2017). *Mathematikaufgaben zur Binnendifferenzierung und Begabtenförderung. 300 Beispiele aus der Sekundarstufe I*. Springer Spektrum.

Ballik, T. (2012). *Mathematik-Olympiade*. Ikon.

Bardy, P. (2007). *Mathematisch begabte Grundschulkinder – Diagnostik und Förderung*. Springer Spektrum.

Bardy, P., & Hrzán, J. (2010). *Aufgaben für kleine Mathematiker mit ausführlichen Lösungen und didaktischen Hinweisen* (3rd ed.). Aulis.

Bauersfeld, H., & Kießwetter, K. (Eds.). (2006). *Wie fördert man mathematisch besonders befähigte Kinder? – Ein Buch aus der Praxis für die Praxis*. Mildenberger.

Benz, C., Peter-Koop, A., & Grüßing, M. (2015). *Frühe mathematische Bildung: Mathematiklernen der Drei- bis Achtjährigen*. Springer Spektrum.

Beutelspacher, A. (2005). *Christian und die Zahlenkünstler – Eine Reise in die wundersame Welt der Mathematik*. Beck.

Beutelspacher, A. (2015). *Kryptologie. Eine Einführung in die Wissenschaft vom Verschlüsseln, Verbergen und Verheimlichen* (10th ed.). Springer Spektrum.

Beutelspacher, A., & Wagner, M. (2010). *Wie man durch eine Postkarte steigt ... und andere mathematische Experimente* (2nd ed.). Herder.

Daems, J., & Smeets, I. (2016). *Mit den Mathemädels durch die Welt*. Springer.

Engel, A. (1998). *Problem-solving strategies*. Springer.

Enzensberger, H. M. (2018). *Der Zahlenteufel. Ein Kopfkissenbuch für alle, die Angst vor der Mathematik haben* (3rd ed.). dtv.

Fritzlar, T. (2013). Mathematische Begabungen im Grundschulalter – Ein Überblick zu aktuellen Fachdidaktischen Forschungsarbeiten. *Mathematica Didacta, 36*, 5–27.

Ganser, B., Schlamp, K., & Tiefenthaler, H. (Eds.). (2010). *Besonders begabte Kinder individuell fördern. Mathematik 2: Bd. 2. Schwerpunkt Arithmetik* (3rd ed.). Auer.

Goldsmith, M. (2013). *So wirst du ein Mathe-Genie*. Dorling Kindersley.

© The Author(s), under exclusive license to Springer Fachmedien Wiesbaden GmbH, part of Springer Nature 2023
S. Schindler-Tschirner, W. Schindler, *Mathematical Stories II - Recursion, Divisibility and Proofs*, essentials,
https://doi.org/10.1007/978-3-658-38611-5

Grüßing, M., & Peter-Koop, A. (2006). *Die Entwicklung mathematischen Denkens in Kindergarten und Grundschule: Beobachten – Fördern – Dokumentieren.* Mildenberger.

Institut für Mathematik der Johannes-Gutenberg-Universität Mainz, Monoid-Redaktion (Ed.). (1981–2019). *Monoid – Mathematikblatt für Mitdenker.* Institut für Mathematik der Johannes-Gutenberg-Universität Mainz, Monoid-Redaktion.

Jainta, P., Andrews, L., Faulhaber, A., Hell, B., Rinsdorf, E., & Streib, C. (2018). *Mathe ist noch mehr. Aufgaben und Lösungen der Fürther Mathematik-Olympiade 2012–2017.* Springer Spektrum.

Käpnick, F. (2014). *Mathematiklernen in der Grundschule.* Springer Spektrum.

Kobr, S., Kobr, U., Kullen, C., & Pütz, B. (2017). *Mathe-Stars 4 – Fit für die fünfte Klasse.* Oldenbourg.

Kopf, Y. (2009). *Mathematik für hochbegabte Kinder: Vertiefende Aufgaben für die 3. Klasse: Kopiervorlagen mit Lösungen.* Brigg.

Kopf, Y. (2010). *Mathematik für hochbegabte Kinder: Vertiefende Aufgaben für die 4. Klasse: Kopiervorlagen mit Lösungen.* Brigg.

Krauthausen, G. (2018). *Einführung in die Mathematikdidaktik – Grundschule* (4th ed.). Springer Spektrum.

Kreß, C. (2004). *Das Thema „Rekursion" im Informatikunterricht. Schriftliche Hausarbeit zur Abschlussprüfung der erweiternden Studien für Lehrer im Fach Informatik.* Eingereicht dem Amt für Lehrerausbildung in Fuldatal. https://arbeitsplattform.bildung.hessen.de/fach/informatik/material/Rekursion.pdf. Accessed November 18, 2018.

Krutetski, V. A. (1968). *The psychology of mathematical abilities in schoolchildren.* Chicago Press.

Krutezki, W. A. (1968). Altersbesonderheiten der Entwicklung mathematischer Fähigkeiten bei Schülern. *Mathematik in der Schule, 8,* 44–58.

Langmann, H.-H., Quaisser, E., & Specht, E. (Eds.). (2016). *Bundeswettbewerb Mathematik: Die schönsten Aufgaben.* Springer Spektrum.

Leiken, R., Koichu, B., & Berman, A. (2009). Mathematical giftedness as a quality of problem solving acts. In R. Leiken et al. (Eds.), *Creativity in mathematics and the education of gifted students* (pp. 115–227). Sense Publishers.

Leuders, T. (2010). *Erlebnis Arithmetik – Zum aktiven Entdecken und selbständigen Erarbeiten.* Springer Spektrum.

Löh, C., Krauss, S., & Kilbertus, N. (Eds.). (2016). *Quod erat knobelandum: Themen, Aufgaben und Lösungen des Schülerzirkels Mathematik der Universität Regensburg.* Springer Spektrum.

Mania, H. (2018). *Gauß: Eine Biographie* (4th ed.). Rowohlt Taschenbuch.

Mathematik-Olympiaden e. V. Rostock (Ed.). (1996–2016). *Die 35. Mathematik-Olympiade 1995/1996 – Die 55. Mathematik-Olympiade 2015/2016.* Hereus.

Mathematik-Olympiaden e. V. Rostock (Ed.). (2010). *Die 49. Mathematik-Olympiade 2009/2010.* Hereus.

Mathematik-Olympiaden e. V. Rostock (Ed.). (2013). *Die Mathematik-Olympiade in der Grundschule. Aufgaben und Lösungen 2005–2013* (2nd ed.). Hereus.

Mathematik-Olympiaden e. V. Rostock (Ed.). (2015). *Die 54. Mathematik-Olympiade 2014/2015.* Hereus.

Mathematik-Olympiaden e. V. Rostock (Ed.). (2017). *Die 56. Mathematik-Olympiade 2016/2017.* Adiant Druck.

Mathematik-Olympiaden e. V. Rostock (Ed.). (2017–2018). *Die 56. Mathematik-Olympiade 2016/2017 – Die 57. Mathematik-Olympiade 2017/2018.* Adiant Druck.

Meier, F. (Ed.). (2003). *Mathe ist cool! Junior. Eine Sammlung mathematischer Probleme.* Cornelsen.

Menzer, H., & Althöfer, I. (2014). *Zahlentheorie und Zahlenspiele: Sieben ausgewählte Themenstellungen* (2nd ed.). De Gruyter Oldenbourg.

Müller, E., & Reeker, H. (Eds.). (2001). *Mathe ist cool! Eine Sammlung mathematischer Probleme.* Cornelsen.

Noack, M., Unger, A., Geretschläger, R., & Stocker, H. (Eds.). (2014). *Mathe mit dem Känguru 4. Die schönsten Aufgaben von 2012 bis 2014.* Hanser.

Nolte, M. (2006). Waben, Sechsecke und Palindrome – Erprobung eines Problemfeldes in unterschiedlichen Aufgabenformaten. In H. Bauersfeld & K. Kießwetter (Eds.), Wie fördert man mathematisch besonders befähigte Kinder? – Ein Buch aus der Praxis für die Praxis (S. 93–112). Mildenberger.

Padberg, F., & Benz, C. (2011). *Didaktik der Arithmetik – Für Lehrerausbildung und Lehrerfortbildung.* Springer Spektrum.

Ruwisch, S., & Peter-Koop, A. (Eds.). (2003). *Gute Aufgaben im Mathematikunterricht der Grundschule.* Mildenberger.

Schiemann, S., & Wöstenfeld, R. (2017). *Die Mathe-Wichtel. Bd. 1. Humorvolle Aufgaben mit Lösungen für mathematisches Entdecken ab der Grundschule* (2nd ed.). Springer Spektrum.

Schiemann, S., & Wöstenfeld, R. (2018). *Die Mathe-Wichtel. Bd. 2. Humorvolle Aufgaben mit Lösungen für mathematisches Entdecken ab der Grundschule* (2nd ed.). Springer Spektrum.

Schindler-Tschirner, S., & Schindler, W. (2019). *Mathematische Geschichten I – Graphen, Spiele und Beweise. Für begabte Schülerinnen und Schüler in der Grundschule.* Springer Spektrum. (English translation (2021): *Mathematical Stories – Graphs, Games and Proofs. For gifted students in primary school.* Springer.)

Steinweg, A. S. (2013). *Algebra in der Grundschule – Muster und Strukturen – Gleichungen – Funktionale Beziehungen.* Springer Spektrum.

Strick, H. K. (2017). *Mathematik ist schön: Anregungen zum Anschauen und Erforschen für Menschen zwischen 9 und 99 Jahren.* Springer Spektrum.

Strick, H. K. (2018). *Mathematik ist wunderschön: Noch mehr Anregungen zum Anschauen und Erforschen für Menschen zwischen 9 und 99 Jahren.* Springer Spektrum.

Verein Fürther Mathematik-Olympiade e. V (Ed.). (2013). *Mathe ist mehr. Aufgaben aus der Fürther Mathematik-Olympiade 2007–2012.* Aulis.

Printed in the United States
by Baker & Taylor Publisher Services